Communication Games

Approaches to Applied Semiotics
5

Mouton de Gruyter
Berlin · New York

Communication Games

The Semiotic Foundation of Culture

by
Eduardo Neiva

Mouton de Gruyter
Berlin · New York

Mouton de Gruyter (formerly Mouton, The Hague)
is a Division of Walter de Gruyter GmbH & Co. KG, Berlin.

♾ Printed on acid-free paper which falls within the guidelines
of the ANSI to ensure permanence and durability.

Library of Congress Cataloging-in-Publication Data

Neiva Júnior, Eduardo.
Communication games : the semiotic foundation of culture / by Eduardo Neiva.
p. cm. — (Approaches to applied semiotics ; 5)
Includes bibliographical references and index.
ISBN 978-3-11-019046-5 (hardcover : alk. paper)
1. Communication and culture. 2. Semiotics. 3. Game theory.
I. Title.
P94.6.N45 2007
302.2—dc22
2007011126

ISBN 978-3-11-019046-5
ISSN 1612-6769

Bibliographic information published by the Deutsche Nationalbibliothek

The Deutsche Nationalbibliothek lists this publication in the Deutsche Nationalbibliografie; detailed bibliographic data are available in the Internet at http://dnb.d-nb.de.

Cover design: Christopher Schneider, Berlin.
Printed in Germany.

To Maria,
Gabriel, Pedro, and Eduardo;
and to the memory of Eduardo,
my father,
with all my love and dreams:
é ancora la sangre oltre la morte

Acknowledgments

Another version of the chapter "Origin" was published in *Culture in the Communication Age*, edited by James Lull (2001, London: Routledge, pages 31-53), as "Rethinking the foundations of culture." I thank not only the publishers for allowing the reprint of this material, but also James Lull for editing the original version of this chapter.

I recognize the help and the encouragement of friends and colleagues. They are Norman Bryson, Arthur Omar, Carlos Cordeiro de Mello, Jacob Klintowitz, Mauro Villar, Fernando Sá, Monica Rector, Keith Moxey, Jair Ferreira dos Santos, Richard Lanigan, Maurício Lyrio, Ronaldo Helal, Traci Crenshaw, Minabere Ibelema, Kevin G. Barnhurst, David Basilico, Jonathan Amsbary, and – once again – James Lull.

The librarians of UAB's Mervyn Sterne Library were immensely helpful in my search for bibliographical sources. I have to thank Eddie Luster, Rebecca Naramore, and Heather Martin for their diligence and attention.

I am grateful to Carlo Romano who was vital in putting the manuscript in shape. I must also thank Jessica Allison for her assistance.

My deepest gratitude goes to Dana Jacobson. She read the whole manuscript, editing it with uncommon understanding, rigorousness, and competence. During many moments of our collaboration, Dana made me clearer to myself. Her literary acumen is only matched by her generosity.

Julio Jeha generously helped in the final stages of writing the manuscript. Without the resort to his skill, and sound advice, the text of *Communication Games* would have been another one.

Contents

Part 3
Individual games

Foreword

Communication Games confronts the assumption that cultures are systems of rules exclusively inherent to human groups. This conception of culture not only gives undue dominance to a narrow type of sign fashioned primarily by conventions – the symbol – thus fostering indifference to the welter of representations active in life, but also brings with it corollaries that have hindered the humanities and the social sciences.

In the chapters ahead, *Communication Games* criticizes aspects of the tradition upon which cultural and semiotic studies have been built, presenting as an alternative to this quandary a conception of semiotics of culture outside of the conventionalist dogma. The book blends the vast arc of Peircean semiotics to the Darwinian understanding that sexual selection is at the core of biological life. After recognizing that sexual selection incessantly creates individual organisms with conflicting interests, *Communication Games* analyzes the positive contribution of game theory to the semiotic study of conflicts and mutual aid as well as to culture and communication.

Communication Games also rejects the repetition *ad nauseam* that cultural conventions establish and determine the identity of social actors. While oversensitive to anything faintly resembling biology, hence accusing whoever is inspired by biological discoveries of determinist and politically reactionary, conventionalists are incapable of admitting that their theories and analyses disseminate a conservative conception of social life. Still, conventionalism is frequently disguised to cloud a special brand of cultural determinism.

Cultural determinism presumes that each social actor is the direct effect of collective thought. The rules circulating in a culture are the means of generating strong cohesion and inescapable enforcement, all under the guise of required social coordination. If that is true, cultural rules precede and determine the exchange of messages in the social arena, shaping simultaneously their production and reception. One may ask: What is wrong with this idea? Indeed, to see cultural rules as the foundation of social life has not always been unattractive. Several decades ago, at the dawn of the anthropological ideology, the drive of this hypothesis came from its seeming capacity of elucidating the undeniable diversity of human cultures. For cultural determinists, human groups were guided by what looked like arbitrary sets of rules, while animals appeared to be the same across the natural world. Cultures determined what was human in opposition to what was part of animal nature. Whoever presented ideas similar to these was reiterating the claim of Jean-Jacques Rousseau (1994), originally presented in 1762, that the natural world is closed to change, where freedom and, by extension, human dignity are more often than not absent.

The finger pointed at biological thought, accusing it with the haste of determinism, was then extended toward anyone who took whatever aspect of biological life as the starting point of his (or her) analysis. According to Rousseau, freedom only existed in contractual relations guiding human interactions. Who could deny that? It is enough to compare neighbouring social groups and the differences created by arbitrary conventions should be more than evident. While that may have been correct superficially in the case of comparing distinct social groups, if one adopts this conception fully, cultural rules are inevitably conjured up as more or less rigid agencies of social enforcement. Paradoxically, if freedom is the defining characteristic of human groups, then something is missing here. Anthropologists, like Clifford Geertz (1973) and Marshall Sahlins (1976), suggested that human cultures could not be compared. Any statement about cultural universality was not only wrong, but also trite. After declaring that there is no such a thing as a universal human nature, Geertz (1973: 49) suggested that this was the most generalizing statement possible about the problem. Each singular human action was nothing but the apparition – faithful or distorted – of the general direction of one's culture. Cultural phenomena are incomparable and should always be thought of in the plural.

Yet, why would human groups be incomparable if our species is so recent in terms of natural evolution? How could the forms of human life be absolutely different from the other mechanisms of biological life? Strangely enough, the voluntarism of cultural determinists mingled with the outrage of religious creationists in their mutual denial of Darwinian biology. Among other ideas, *Communication Games* asserts that each human society has many commonalities, not only with other social human groups, but also with animal societies. The divergence of human and animal sociability is definitely one of degree, not of kind.

About time

The chasm between the humanities and biological thought was not based on facts, but on theoretical bias. Nowhere is this prejudice more evident than in the way that the complexity of natural and evolutionary time has been conceived by the humanities and the social sciences.

Much before Rousseau's separation of culture and nature, the temporality of human life was deemed altogether unlike natural time. With Plato, historical flow was perceived as degradation and loss, a process of decadence regretfully akin to what occurs in natural life. Plato professed that the proper philosophical approach to the timeless realm of forms demanded the abandonment of what is transiently natural and human, requesting the attention toward a world of eternal permanencies; although, from another perspective, it could be equally possible to consider the flux of time as the expression of betterment, freedom, and innovation. Nonetheless, whatever the final answer to the evaluation of human history for the humanities, the questions around the topic of time developed from an intuition that was both unquestioned and mistaken: the tendency to state uncritically the disconnection between the culturally learned and the biologically inherited. That happens even with thinkers – like Martin Heidegger (1996) – who criticize the post-Platonic tradition and yet segregate natural and human temporality. It is only logical for Heidegger to attack what seems to him the deleterious representation of time in terms of natural events. Like an enraged oracle, he preaches against the understanding of time in terms of things and objects, which, according to him, entails an irretrievable deviance from what defines humanity.

The temporality of human beings could not be that which is identical to the sequential intervals of hours, minutes, and seconds, represented by the paced movement of the hands of the clock. For Heidegger, the sacred dimension of human temporality was beyond the limits of things. To reduce time to the sequence of clocks was a detour from the truly ontological dimension of human existence. Natural time was one thing; human history, another. At the heart of Heideggerian statements lurked the ominous threat of separation between natural and human time.

Such a conception of natural time opposed to strictly human time was nothing but illusion. Darwin's evolutionist ideas showed that natural bodies, whether human or not, sustained a shared temporal inscription. As never seen before Darwin, the separation of nature and culture was pinned down as an arbitrary and capricious frontier. So, if one compared a fruit fly, a drosophila, with the genetic human structure mapped by the *Genome Project*, the result was the recognition of striking molecular similarities between the organisms, when, in the remote past, one expected parallelism of species, and no convergence at all between them. It is, therefore, not surprising that some of the theoretical constructs of the humanities prevalent one century ago were vulnerable to refutation, while the Darwinian intuition bloomed as the procedures of life were scientifically revealed. That took place despite one of the great intellectual misfortunes of the nineteenth century: evolutionary theory developed in the absence of Darwin's reading of Mendel's monograph on genetics, which would have helped Darwinian theory to cope better with the phenomenon of individual variation of sexual organisms and evolutionary change (Mayr 1990: 109). Yet, in any case, Darwin's groundbreaking book of 1859 – *On the Origin of Species* (1979) – touched a striking chord, proposing that life is marked by common descent with modification.

Although it took some time to be established as the principle of organic life, the selective and central unit of evolutionary explanation became the competitive individual organism, and on a deeper level the replicators, the genes, but never the totality of the species. The unit of evolutionary selection – whether under the form of survival or replication – must be defined in terms of singularity. Group selection is an ineffective and unwarranted explanation, being just too slow as a discriminating method when compared to the choice of individuals; it is from individuals that life radiates. That is why each singular living organism bears in its body the history of their ancestors and, for the same reason, becomes a potentially evolutionary new beginning. Looking forward, one sees individual organisms branching out into other organisms that are modifications of themselves; but looking backward, in the direction of their origin, what one sees is unity.

Biological time has two simultaneous directions; temporal flow forks out into variations that spring from a common origin. The living process is both diversity and commonality. This is the most radical critique made to the traditional ideas of time, history, and memory prevailing in the humanities. More like the steps of a hopscotch jump, evolutionary history escapes the grid of being a straight linear progress. The flow of time is progressively drawn and redrawn, going forward and backward. Through the action of the parts, in their extreme individuality, it is possible to cancel out what was whole until then. Without giving any precedence to human nature, all natural life is defined by evolving liberty.

Although in different degrees, the natural and the human worlds are open universes. During the unravelling of time, historical development specifies ceaseless renovation, indeed endless transformation whether for good or worse. In evolutionary theory, the expectation of collective progress simply does not make sense. Individual success far outweighs the improvement of a group of organisms. Nonetheless, the course of time reveals that the products of human culture are always considerably and inevitably modified; otherwise, there would be no diversity of human practices and social institutions across the world. Considering the notion of common descent with modification, it is inevitable to infer that the same occurs, though in a slower pace, within the realm of nature. We are not completely like other living organisms, and yet all living organisms have in their bodies not just the history of their species, but of all forms of life. DNA is, in its structure, the recombination of the same basic molecular elements for all living things; our brain is marked, for example, by the heritage of the brain of reptiles. If each living organism carries with it the total history of biological life, it is impossible to accept the holistic notion of anything prior to individuals in interaction. It is a cognitive absurdity and a factual inconsequence to uphold theoretical interpretations that are nothing but the products of the totalizing and collectivist ideologies that ruled the humanities and the social sciences since the last century.

The historical persistence of total ideologies

Either as historical interpretation or as sociology of knowledge, total ideologies have postulated that external causes and consensual social interests act upon individuals with such strength that not only themselves, but the ideas generated during the course of history are passive reflections of the moment and the society where those social actors lived.[1]

Total ideologies suppose that forces larger than the individuals mould, consciously or not, what is intellectually produced in a specific moment of history or within a culture. It is a unifying notion, definitely collectivist, presented in a variety of forms, whether as the spirit of an age, a class, a nation, a race, a social group, or a culture. In any case, interpretations based on the suggestion of total ideologies display a holistic image of social life.

The idea that cultural messages should be evaluated considering what is external to them, a notion cherished by both Hegel and Marx, was directly formulated in the middle of the nineteenth century at the same time that Darwin was brooding about the individualistic destiny of living organisms. In opposing and yet stunningly similar ways, Hegel and Marx conceived human history as progression. For Hegel (1949: 801-808), history showed its progress through the rational overcoming of contrary historical moments, a complex blend of simultaneous conservation and surmounting, expressed in one German word: *Aufhebung*. In Hegelian terms, the whole of history is the true expression of an absolute spirit, hovering above time and conducting it. History is thus the conciliation of contraries and the achievement of unity moving in a predetermined direction. While Marx also thought of human history as a progression, for him, it was nonsensical to think that a rarefied and idealized spirit could be the dynamo of history. The Marxist vision of history presumes the inevitably violent resolution of the economic contradictions that set human beings in opposing camps. In other words, history was nothing but the struggle for economic survival and domination (Marx and Engels 1848).

For that reason, in *The German Ideology* (1998), Marx and Engels originally indicated in 1846 how Hegel's idealistic philosophy and its development were the expression, in disguise, of the ideological tactics and interests of the ruling class; yet, Marx himself could not be excused from sponsoring a theory laden with an equally total ideology. Marx effectively argued that it was through the viewpoint of class interest that one should understand the meaning of the philosophical ideas of his time, as in the case of Max Stirner's (1805–1852) anarchic individualism, or of Ludwig Feuerbach's (1804–1872) rosy hope that religion was the tool for the improvement of humanity. The difference with Marx was that he thought that he was taking the active side of the working class and saw himself not as another philosophical conjurer, but as an involved, conscious participant in the achievement of human freedom.

Without any great reverence for Hegel's grand hypothesis that history rationally overcomes social contradictions by itself, Marx proposed the alternative of actual political activism through which it is feasible to forge a classless society without parts inevitably fighting each other. At the end of the struggle, social life would be definitely transformed, and it would rest in a utopian communist whole, free from private property and divisive economic oppression. Marx could proudly declare to have turned Hegel upside down. Indeed, Marx indicted Hegel's interpretation of history, revealing how the Hegelian interpretation ought to be seen from the perspective of the sociology of knowledge. To comprehend Hegel, it was indispensable to answer a fundamentally political and sociological question: What was the true, although hidden, social interest of Hegelian idealism? Marx's sociology of knowledge accused Hegel of being an effect of social drives larger than himself. Moreover, Hegel's idealism was just part of a ploy contrived to affirm conservative political goals; but was Marx very different from Hegel? Hegel's passive conventionalism was supposedly destroyed and replaced when Marx got into the scene; yet what Marx did was limited: he overturned one set of conventions and substituted another. As the dictatorship of the proletariat crystallized, new rules, the precise inversion of bourgeois conventions, should be woven into the fabric of society. Conventionalism remained an untouched common trace of Hegel and Marx: Hegelian speculative philosophy of history simply mutated into Marx's total ideology of economic collectivism.

At least in one aspect, the Marxist sociology of knowledge should be taken seriously. It would be a naïve abstraction to think that human knowledge manifests itself in a historical and social void and that the impact of ideas and their practical effects are devoid of the concrete interests of human beings competing with another for resources. Ideas thrive neither in ivory towers nor in arid lawns. It is certainly simplistic to presume that human theories are immune from the appropriation and the use of social actors, who will be stirred to defend their immediate interests. That is a possibility that should not be dismissed; the social and historical character of human intellectual activity is undeniable; but it is something else to postulate that all ideas simply reflect passively and uncritically the class interests and the social prejudices of their creators, or then of the ruling social strata. If that were true, it would be impossible to criticize and revise any set of notions. All forms of knowledge and culture would remain stagnant; no alternative would ever be presented to dominant notions. Each system of ideas would exist in an everlasting world of its own, closed in itself, and unchanged, living in a bell jar.

This type of historical and cultural holism is an astonishingly robust trend that cuts across several conceptions of culture and knowledge. It goes from Hegel's primacy of a collective spirit of an age and his speculation that history unfolds continuously, rationally, and inexorably in the direction of what he considered to be the substantial will of the spirit of the world, indeed a lofty idea of liberty, to the conception of paradigms and *epistemes* organized in disconnected layers, nonetheless presiding over the individualities of cognitive social actors. This is the idea of history that both Thomas Kuhn (1962) and Michel Foucault (1972) sponsored after the triumphant assumptions of Marxist thought and Freudian psychoanalysis.

To adopt merely the unqualified perspective of the individual without a radical reformulation of conventionalist, historicist, and holistic working hypotheses is no guarantee that the strictures of total ideologies will be kept at bay. Consider the case of Freudian psychoanalysis that seems to foster an understanding of the individual from the radical standpoint of personal destiny and history and yet exudes conventionalist holism. Freud may have become an old-fashioned Victorian gentleman, holding fast to an umbrella, but his ideas have been vitalized by Jacques Lacan (1977). The contemporary Lacanian reading of Freudian theory relies heavily on the critical rejection of Freud's early, although awkward, biological tenets. Again, one finds in Lacan's neo-Freudian interpretation of the psychoanalytic system of ideas the recurrent division of biological nature and human. For Lacan, the social, the symbolic, and the conventional orders shape the psychological identity of the individual. The child will necessarily have the personality formed as a direct result of the fear of symbolic castration. Sexual desires of the child ought to be repressed, and then the individual is fully socialized. The cultural order reigns over the individual. In a cunning expression, impossible to be meaningful perhaps in few languages other than French, Jacques Lacan stated that the awareness of the social and cultural order was dependent upon the recognition of *Le Nom du Père* (The Name [No] of the Father), which is a tight pun merging the ideas of the name (*nom*) and the cultural no (*non*) of the Father. In the French language, the words *nom* and *non* are almost undistinguishable phonetically. Moreover, Lacan is not talking of actual fathers, but of the symbolic presence of *The Father*. Actual fathers are merely the lieutenants of the overall and indispensable rule of culture, without which society could not even exist. An individual absolutely released from the strictures of culture is either an abstraction or a psychotic personality, certainly a prey to unfettered desires and representations. Hanging between the extremes of psychosis or neurosis, individuals suffer from either lack or excess of order. In any case, they are wholly dependent on the action of culture.[2]

A misleading alternative to total ideologies

When a total and synchronic system subordinates the exchange of signs in society, the effect is that the individuals that trade them have little or no import at all. Each interaction between individuals just reflects and is subjected to the whole allowing and establishing the way in which social actors deal with one another. If that is so, the effective agency of social life is never the actor, but the culture that creates and anticipates social interaction; social facts unfold from other social facts that are irreducible to the personal psychology of each individual.

For that reason, in *Les Règles du Méthode Sociologique*, published in 1895, Émile Durkheim (1960: 102) declared that the social whole is not identical to the sum of its parts, and that this kind of whole is altogether different: its properties are not equal to the parts that compose it. He expressed clearly that the synchronic dimensions of the social facts are the major trait of collective representations. Performed signs are simply guided by and submitted to social rules.

Against other social thinkers of the nineteenth century, who were – like Hegel, Marx, Freud, and even Darwin – entranced by a historical perspective, Durkheim dismissed the dominance of historical progress. For him, what truly counted was the identification of current social laws at work over the minds of actors, while for some of his historicist contemporaries the past seemed to carry more relevance than the present. In Durkheimean terms, the emphasis should be on the understanding of the immediate causes of specific social states. Historical analyses were just too far way from the demands of sociological analysis. The comprehension of what creates a steady state of social affairs was infinitely more important than the ripples of historical causation. The French Sociological School owes to Durkheim the notion that one should discard what is diachronic and favor unequivocally the analysis of synchronic states. That is the condition to produce a sociological interpretation. It could not be otherwise, for that is how social actors in fact live. Social groups are structured around the exclusive action of collective representations guiding individual choices.

While history and tradition may in part explain why this specific historical moment came to be the way it is, what happens nowadays is always more central. Social actors behave in their specific fashion as result of the coercion of social conventions that are always aligned according to the consensual directions of the social whole. The submission to a collective view is a condition for the individual to exist and remain within the group. Lacan's evaluation of the predominance of the symbolic order over the individual psyche is a redressing of the Durkheimean conception of society. The stability of social experience depends on the coercion of its system of rules and conventions. Conventional rules are directly mirrored in the signs that circulate in societies, guiding their meaning and distinctive form.

It is from his reading of Durkheimean sociology, seen in the sprinkling of sociological terms sparkling over the textual surface of the *Cours de Linguistique Générale* (1916), a collection of class notes from his disciples Bally and Sechehaye, that Ferdinand de Saussure[3] derived his fugitive vision of a general science of how signs exist socially, which he called *sémiologie*. According to Bally and Sechehaye's reconstruction of three Saussurian courses on general linguistics, Saussure alleged that each exchanged sign in human interaction is the result of actual rules that exist in synchronic and systematic solidarity: the rules are formative structures, the determinant element in the production of human communication.

Concerning the element time, Saussure's relation to the Durkheimean accent on the present is more complex than a simple appropriation (Yong-Hoi 2002). For Saussure, time still mattered; and that meant that the linguistic dimension was not achronic or static. The interest should be on what remained stable as time passed. This stability reveals, by default, what is to be given precedence to; it is a dense notion of how the actual system of linguistic rules is present in the mind of the speaker.

The object of linguistic studies could never be the individual speech (*parole*) of speakers, but the system of conventional rules that precede and determine individual linguistic acts. The deep kinship of Durkheimean sociology and Saussurian linguistics was noticed a long time ago. Doroszewski (1933) pointed out that when Saussure labelled *langue* as a shared set of rules, more vital than actual speech, he was echoing Durkheim's distinctions of *représentations individuelles* and *représentations collectives.* Collective representations create social facts. For Saussure, language-structures (*langues*) are also within collective conscience; the social nature of languages places them beyond individuals. Language is a set of rules bestowed on the social actors. The essential trace of any language-structure is "its imposition upon the individual through collective usage" (Doroszewski 1933: 90). The holistic assumptions of the French Sociological School are obviously visible in Saussure. In a far more recent evaluation of the Saussurian legacy for both linguistics and semiotics, Bouissac (2004a and 2004b) noted that language-structure (*langue*) was a daunting concept for Saussure. *Langue* was a set of differential terms, founded on arbitrary conventions that were not conscious in the mind of the speakers. If the language-structure was a set of conventions, it was obviously learned; nonetheless, the process of learning had to be forgotten; otherwise, it could not be so automatic. Moreover, it was forgotten that it was forgotten; only then could it function as a blind force over the speaker. Qualifying *langue* as "a contract without contractants," Bouissac (2004a: 11) declares it a paradox. However, according to Durkheim's sociological program of research, the idea is quite logical: individuals are of no importance; what is relevant is the action of the social contract, the forceful push of conventions.

Many who read Saussure, oblivious of his Durkheimean pedigree, ended up misunderstanding what he said and were lead to believe that the theory of signs in the *Cours de Linguistique Générale* inaugurated a relativistic, and supposedly libertarian, almost post-modernist view of social life; but how could that possibly be so if collective representations imposed normative conventions that coerced individual consciousness? This truly bizarre interpretation is suggested when one reads Saussure's chapter on the nature of the linguistic sign out of context and stumbles upon the apparently stunning argument about the arbitrariness of the linguistic sign. The notion of arbitrary linguistic signs is problematic: Benveniste (1966: 49–55) showed the imprecision of the argument, noting that the sign may be nonmotivated, but it had to be perceived as a necessary representation by the speakers of a language. Saussure's emphasis on an arbitrary linguistic sign is an argument similar to the incommensurability of human cultures. Superficially, all languages seem to be incapable of being compared; they appear to be hedges that cannot be easily crossed, or mirrors fatefully opaque to one another. Forgetful that young children learn languages without formal coaching at all, at the mere exposition to distinct linguistic means of expression, crude readers of Saussure were satisfied with the answer that signs are recognized by their users for the positive reason that they were produced from systems of rules commonly shared within the social group. Only conventions counted; and they could change: they were arbitrary. Rousseau's adamant contractual image of social practices entered again through the backdoor of this reading.

The effects of Durkheim over Saussure went well beyond the disciplinary frontiers of sociology and linguistic. After Saussure's adoption of Durkheimean tenets, grasping the meaning of practices and modes of human interaction could be reduced to the detection of the codes that rule them. With Claude Lévi-Strauss (1945 and 1966), who perceived with enthusiasm the viable merging of Durkheim and Saussure, anthropology did more than turn out staple descriptions of archaic societies; it began to be enticed by the discussion that its intellectual goal could well be to recognize the normative codes underlining social representations.

Even history was called to the task of identifying the rules and dominant conventions of an age, presuming that, while historical interpretation is always inferential and hypothetical, for the past is only accessible through signs that are left out to the present, each historical period is uniform as result of its common rules of formation. These were the assumptions of Carlo Ginzburg. Historical signs require interpretation, but what kind of interpretation constitutes history? For Ginzburg, historiography was the product of an exclusively conventional and strictly symbolic theory of signs. That is very evident in two studies of his (Ginzburg 1989): "The high and the low: the theme of forbidden knowledge in the sixteenth and the seventeenth centuries" and "Titian, Ovid, and the sixteenth century codes for erotic illustration." In the first study, the thrust of Ginzburg's analysis is the recognition of a matrix of oppositions (high as against low) quite close to the most typical structuralist analysis; in the second text, the idea of code is a first cousin of the previous enchantment with underlying oppositional structure. In both studies of Ginzburg, the historian is making a conventionalist anthropology of the past. Conventionalism seemed to be all over the place.

After such a long time, more than one century later, the idea of social coercion that appeared so vital for Durkheim was moderated, and frequently overlooked, but the Durkheimean roots of conventionalist semiotics have not been pulled out. A vast array of contemporary intellectual attempts – media studies, gender studies, post-colonial studies, for instance – that receive the common tag of cultural studies still suppose the ruling of collective representations over the individual. Cultural studies rely implicitly on the assumption that individuals are the upshot of collective representations, themselves strong enough to forge the identity of individuals and to construct the foundations of societies. Such a claim of social construction is senseless without the belief in the force of conventions over social actors. It is nowadays unfashionable to speak of the spirit of an age and its vision of the world, of class dominance, or even the formative function of the unconscious, although that last topic to a much lesser degree. Never discussed openly, the present trends of the humanities are couched in an unambiguous conventionalist theory of signs. It is urgent to nullify the certainty that rules and conventions create social practices and their actors through the mere deviance or following of predetermined conditions of sign production.

Game, the alternative

The presence of the word *game* in the title of this book and in its three major sections – "Canonical Games," "Ancestral Games," and "Individual Games" – is reason enough to present the prompt definition of the concept of game. The concept of game that it embraces was presented initially in John von Neumann and Oskar Morgenstern's *Theory of Games and Economic Behaviour* (1944). Following their insight, game is much more than what is regularly understood from the ordinary use of the word. For *Communication Games*, game is defined as a situation in which individual interests collide. The individuals involved in the game – the players – interact with one another in such a way that the choices concerning the course of their actions decide the outcome of the clash, translated unequivocally as the result of who wins and who loses. According to von Neumann's early formula, clothed by the now popular term zero-sum game, the conflict of interest is plainly represented by the victory of one player over another.

A particularly exciting feature of von Neumann's massive mathematical demonstration was the clue that game theory will always suggest the proper and most efficient way of playing in a variety of circumstances: so, game theory is not to be restricted to the analyses of recreational interactions. Game theory deals with a bewildering number of situations; it covers conflict of interest such as auctions, haggling, bargaining, the effects of the buying and the selling of real estate, the strategies during the courtship of a seducer facing a reluctant mate – as is so common in the love game – notwithstanding labour negotiations, political debates between contenders in search of votes, fluctuation of values shown in the stock market, skirmishes between social groups or nations, and diplomatic minuets. In all of the situations just mentioned and in many more across human societies, the strategies and plans of the players in conflict are developed taking into account what the opponent might do. Game theory is a strictly relational theory; to understand it fully, one must always have an opponent's most effective strategy in mind.

Communication Games argues that among all theoretical models produced in the social sciences, game theory is specially suited to the advancement of communication studies. The reason should be evident: like game theory, communication circumstances must always be at least the interaction of two elements of the process: one, a sender; the other, a receiver, both involved in the exchange of signs. With less than this dual interaction of sender and receiver, connected by the circulation of signs, communication could not exist.

Game theory is at the confluence of communication and semiotics. The match is perfect: their cognitive frame is noticeably similar, although the remote origins of game theory go back not to rhetoric or any other semiotic problem but to an economic analysis: A. A. Cournot's (1960) study of duopolies published in 1838. Cournot tried to comprehend mathematically what occurs when two competitive economic agents – two players – face each other with the mutual goal of driving the other out of business and wishing to dominate the whole market. Almost simultaneously, the other contribution to the awareness that led to game theory was Charles Darwin's discussion of the puzzle of the equal sex ratio in nature: Why is it that one sex – male or female – does not override the other, simply eliminating it? The outcome may appear to be different, but the issue is the same: in Cournot's model, the result is the defeat of the opponent, whereas in Darwin's the competition ends in equilibrium. Economy and biology deal with independent individual agents struggling with one another for economic or biological survival.

The question remains: How does this relate to communication studies? Game theory is particularly relevant for communication because both players in conflict will be trading information as they interact with each other. From the information conveyed or gathered, a strategy – which is a complete plan of action – will be established; and from the implementation of the strategy, the outcome of the game emerges. Just considering this, it is easy to imagine that the perceptions of game theory constitute a precious source to figure out how messages can and should be made. Voluntarily or not, the semiotic information conveyed by the exchange of signs determines the success or the failure of the player in the game.

Sex, selection, and culture

For communication and semiotic studies, more important than the conclusions of economic theory is the biology of sex. As will be seen in "Sex, Signals," a chapter in the section *Ancestral Games*, sex begets individual organisms that will compete with one another for replication. The destiny of sexual beings is relentless competition. If one male organism mates successfully with a female, several other males are inevitably excluded from procreative success until the female can once again procreate. The rule is rather drastic: if one succeeds, others will necessarily fail. That is an obvious zero-sum game, and it explains the insistence of fierce and never-ending competition among sexual beings. One important aspect of all this cannot be denied: procreative sex is competition between males and cooperation between a male and a female. As masturbation breeds no offspring and often demands reveries of other players, sex is never a completely solitary affair: procreative sex takes two, and it also takes communication, as the mates converge in their joint interest through the exchange of semiotic information.

The main line of reasoning of *Communication Games* demonstrates that cultures and communicative interactions are not simply means of integrating social actors to a pre-existing group. The performance of an individual organism within a culture provides fitness indicators for sexual selection. Cultural performance is a manner of segregating social actors both within and outside of the group. This book postulates that the analyses of culture should begin with the perspective of competition to move then to understand the cooperative webs woven in social interaction. *Communication Games* aims at subverting the traditional assumption of cultural and anthropological disciplines, contending that conflict and risk cannot be avoided in communitarian life; in fact, they are crucial elements in social living. Human beings can try to warn against risk and conflict and even dream of their extinction, but it is impossible to escape biological imperatives bequeathed. It is only thinking of life and all of its manifestations that one can fully blend the dimensions of competition and cooperation constantly at work in cultural experiences.

For a feature so persistent in the great majority of living organisms, sex has not been served well in the humanities. While anthropology talks of kinship systems, and cultural studies hammers on the positively peculiar notion of gender – strange because "gender" is really a pronominal term, used primarily with the ideological purpose of reinforcing the conventionalist conjectures of social constructionism – *Communication Games* underscores the raw mechanisms of natural life and of sexual interests as the foundations of human culture. While gender studies culturalize sex, *Communication Games* sexualizes culture.

Although the current idiom sets competition and cooperation as mutually excluding facets of human interaction, this is not the case at all, as long as the problem is examined at close range. Without eliminating competition, indeed because of competition, cooperation evolves as social players interact. In the natural world, cooperation is a widely established. Death is to be announced when the organs in the body stop cooperating and following the signals circulating across the organism.

Cooperation works always to the best interest of selfish individuals. Cooperation persists in the natural and the human worlds not because of altruism and self-sacrifice, but as the direct result of self-interest. There is no contradiction here: How could cooperation be effective if it were contrary to the interests of the players involved in it? In that case, cooperation would have been picked out from life. The thought that that the actions of individuals within a culture are above all directed toward the cohesion of the group is a chimera. The concept of culture should be sexualized. The individual social actor sees the group not as a lofty totality, but a sexual and thus genetic pool, a stage to display talents in the hope of being sexually selected. Inevitably, individuals serving their immediate interests secure the cooperative ground of human society and culture.

Contrary to individual sexual selection, group selection depends on an intricate and cumbersome device with dire social consequences. If the group must curb and ideally eradicate selfishness, the action of enforcers, totally released from selfishness, would be indispensable. What is the Durkheimean notion of collective representation if not of an abstract enforcer? How is it possible to abolish selfishness in the case of sexual beings? Furthermore, to make sure that these enforcers do their job unselfishly, it is again mandatory that another set of enforcers supervise and eventually enforce the principle of unselfishness. That requires more enforcers, and so on infinitely. Social life is turned into a hell of mirrors. Totalitarianism prowls around collectivistic conceptions of societies.

To avoid accusations of totalitarian collectivism, social theorists would eventually refine culture, making it into the link that holds individuals together. Culture was then reconceived as a means of coordinating selfish actors. According to Durkheim's thesis, the whole curbs the selfishness of the parts. Social submission is accepted in the name of the common goal of achieving social harmony. At this point, coordination is fused with cultural conventions that ought to be constantly and recurrently communicated to social actors so that the group can maintain its integrity. This operational hypothesis is now a dogma in the humanities and the social sciences.

The long road to the canonical conception of culture

The section "Canonical Games" has as its goal the detecting of how the fusion of convention, culture, and coordination was generated. The chapters in "Canonical Games" are all historical and hermeneutical. Their purpose is to show how the current idea of culture took shape. The rise of the contemporary conception of culture came slowly, naturalizing itself along the way, to be finally seen as a force capable of creating simultaneously the character of the group and its members. Returning to the circumstance of its inception, it is possible to perceive how the notion of culture and its attributes were built. To do that, one must survey the historical landmarks forming the concept of culture. Layer was laid upon layer to construct a concept that is presently comprised of complementing sides, just like a rotating prism. Culture is conceived canonically as an arbitrary means of integrating individuals, arbitrary because of its variation from group to group; and for that reason culture turns out to be the device that grants cohesion and stability to the social whole. The cultural system is experienced as a dominant body of ideas within a group, furnishing the principles of its social contract and thus capable of outlining and transforming the individual will of its members. All such harmonized aspects could not have come in a flash.

Even if Thucydides did not use the term "culture" directly in his rendition of Pericles' funeral oration, delivered at the end of the first year of the disastrous war between the Athenians and the Spartans, it is patent that the greatest Athenian political leader is drawing on an idea quite close to our current usage when justifying his unsuccessful war strategy. Pericles mentions the idealized Greek notion of *paideia*, which is an overloaded term covering the contemporary nuances of culture, civilization, education, tradition, formation, and even literature (Jaeger 1939). More important than the term *paideia* is the manner in which Pericles manipulates and converts this idea, as is shown in the chapter "Conflict" of *Communication Games*. Of course, the Greek word cannot be exactly what is now defined as culture, a phenomenon universally ascribed to human groups strewn over the world.[4] Nonetheless, it is acceptable to compare Pericles' usage of *paideia* with our notion of culture, if we recognize a substantial distinction: in Pericles' funeral oration, culture was not an explanatory concept but an argument to rationalize his concerted strategy quite at odds with the Greek tradition of individual heroes, often fighting without a rigid plan, in search of individual military glory. Like Themistocles before him, Pericles had offered a plan to the Athenians. The city took up his idea, but the plan was beginning to show its shortcomings. Pericles' use of *paideia* was nothing but an argument in the theatre of war; yet this is the first time that the idea of a collective trait – the Athenian culture – is seen as being responsible for structuring the mode of life of the members of a group. In Thucydides' historical record, one finds the most ancient use of the idea of culture as well the suggestion of a direct causal explanation for war based on cultural differences.

Accordingly, the chapter "Conflict" of *Communication Games* should also be seen as a refutation of the unrelenting attribution of cultural causes to conflict between groups. The idea is certainly explicit in passages of Thucydides' narrative and in those of Augustine of Hippo, present more recently in the argument of Marshall Sahlins (1977) against the reduction of human cultures to their biological foundations. The same contention is present in Samuel P. Huntington's announcement of the clash of civilizations. In *De Civitate Dei* (*On the City of God*), Augustine contrasts the heavenly city of Christians with the mundane dwelling of pagans: each city is ruled by specific cultural principles that exclude one another. Much later, having in mind Rousseau's conclusion that wars are always the result of states or collective wholes fighting among themselves, Sahlins (1977: 8) professes that warfare is not to be seen from the viewpoint of belligerent individuals, but always from the recognition that such conflicts exist primarily between socially and culturally constituted polities. Huntington (1993 and 1996) thinks along similar lines in a world whose dominant trait is the presence of global policies far more intricate and convoluted than the ones that were part of Augustine's concerns. Trying to understand the effect of today's new global order, Huntington probes into the past to make a prediction. For him, until now, Western wars went through several stages: at first between princes and monarchs, then between nations, and later between ideologies. With the end of the Cold War, the scenario of global conflicts changed. What we see nowadays are clashes of civilizations. Civilizations are the highest cultural groupings: they grant identity to human beings. Huntington is clear in his formulation: cultural elements outline a civilization with precise notions of socialization, economic exchange, religiosity, and political practices.

Huntington's hypothesis is simplistic, but nonetheless apparently confirmed four years later after the publication of his ideas in book form by the attacks of Islamic terrorists on New York's Twin Towers and the Pentagon. At that moment, it seemed that Huntington's observations had the unquestioned value of a confirmed oracle; but there are greater subtleties in wars that go far beyond mere cultural differences. The Peloponnesian War, which spread itself for such a long time and with such devastating consequences, falsifies at least one of Huntington's claims about the degree of violence in conflicts within the same civilization. For him, in the same cultural whole, the clash should be less intense and more controlled than the ones occurring between different civilizations. Nothing like that happened in the years of conflict between the Greeks of Sparta and Athens. The historical examination, presented in "Conflict," offers an alternative to extremely general conclusions linking directly conflicts and cultures. War is far more complex than narrow cultural interpretations presume.

The end of the excessively long war of Athens with Sparta is present in Plato's indictment of democratic culture. Democracy unleashes perilous social forces, allowing futile and vicious discussions of ignorant citizens. Such individuals should never guide the city; they should be under surveillance and guidance. In a democracy, citizens think that they govern themselves, just as alcoholics imagine having self-control, when they do not have it (Pradeau 2004). Individual freedom is a wine much too strong to be taken without coordination and control; adequate conventions dilute the liberty of individuals. The chapter "Coordination" indicates the role of Plato in establishing conventionalism as a central notion in social and political thought.

In Augustine's *Confessions,* the war occurs within the individual consciousness. "Contract" is set around a crucial moment of Augustine's life. He had converted to Christianity and was baptized. He was waiting in the port of Ostia for a naval blockade to be lifted so that he could return to his native North Africa. Talking to his mother, Monnica, Augustine undergoes a mystical experience that confirms the righteousness of his religious option for Christianity. The serene and peaceful ecstasy of Ostia stands in stark contrast with the previous turbulence of his conversion. At that moment, in a garden of Milan, Augustine discovered that he ought to surrender his will to God, and thus renounce his sexual life. Augustine had been given a sign when he picked up and read an epistle of Paul to the Romans. The key to a sign is never the sign itself but a set of values that gives meaning to it. Augustine's conversion coincides totally with his acceptance of the Biblical canon. He not only submitted his inmost will to the cultural rules of ascetic Christianity, but also fervently willed to understand the word of God precisely and faithfully. If before his conversion Augustine was wrecked with doubts, now – after the mystical vision of Ostia – he was sure that it was the correct option, despite all the previous fluctuations of his heart. The mystical experience of Ostia, however, could not be transmitted to anyone; its intensity was lost in even the most accomplished literary narrative. He had received a sacramental sign and should transmit it to other Christians. That was his mission, and he tried to accomplish it at first as a Catholic thinker who wrote several books in the retreat of Cassiciacum, just after his baptism, and later as the Bishop of Hippo.

Even though Augustine worked with a theological vocabulary, he sketched the principles of what turned to be an influential semiotic theory, one that subsumes the material sign to a well-defined set of rules. Doing that, he provided the means for establishing what should orient the ones who aspired to live in the society of the heavenly city. A biding contract extends from the Divine core solidifying, like never before, the whole cultural heritage of the group. With Augustine, the social contract is not anymore an affair between the members of a community: it is an unbreakable bond with God, the infallible and all-powerful enforcer. In nobody else, not even in Plato, whose goal of coordinating social actors drives them away from their base and selfish interests, which ought to be subordinated to a timeless world of ideas, the renunciation of the natural order is as absolute as it is in Augustine. He moved from a universe ridden with changing and unstable signs to achieve repose in the Scriptures that codify God's word. Cultural codes – certainly superior to the transience of material signs –

have to be wholly segregated; they are transformed into an autonomous and self-sufficient order with no natural foundation. Apparently unbeknownst to many anthropologists – like Geertz, Sahlins, and Lévi-Strauss – this idea of culture bounced in a straight line from Augustine's sacramental theology.

The common descent of nature and culture

With different arguments, the chapters "Origin" and "Sex, Signals" seek to write off the prejudice that human societies are a radical break from the natural world. If, on one hand, it is impossible to uphold this assessment remembering today's overhaul of cultural values under the unyielding sway of global communication, on the other hand, the premise of human cultures radically autonomous from each other is unsustainable from the viewpoint of Charles Darwin's revolutionary understanding of the biological processes. Life is not superficial diversity, but the growth of a deep common origin.

"Origin" shows that the same features of human cultures are at work in the animal world. Do they have groups like ours? There are numerous instances of animal societies. Do they communicate through the exchange of signs? They do it as we do. Did they develop modes of cooperation in any way similar to ours? That is obviously true. Have they the capacity of organizing their relationships around conventions, as we seem to do? About that, take, for example, the issue of animal conflict and notice that animals struggling with another of the same species could chose from three basic strategies: outright escalation of fight, threat displays, and conventional tactics. Yet evidence is overwhelming that, in the animal world, evolutionary preference is given repeatedly to the combination of threat displays and ritualised fighting. In the case of deer, fish, or snake, among so many others, animal fighting consistently follows conventions. Conventions cannot be what divide humans and animals. Culture is an integral part of the natural world (Bonner 1980; Mesoudi, Whiten, and Laland 2004; Laland and Hoppit 2003; Whiten, Horner, and Marshall-Pescini 2003).

Methodologically, in both chapters, the individual surfaces as the privileged unit of biological life. If that is the case in the natural world, why should it be different among human beings? The riff between nature and culture is senseless without the Augustinian theological conviction that a Divine Maker created human beings in His image and similitude and thus carved an unbridgeable abyss between humanity and other forms of life.

Without such a conjecture, how can one accept the autonomy of human culture in relation to the natural world? While it is true that Augustinian tenets devoid of his theological underpinnings are just a pipe dream, the autonomy of cultures proliferated through the influence of Rousseau. The chapter "Origin" provides answers to the question of how societies emerge without using Rousseau's version of social contract and binding conventions.

The enigma of sex haunts "Sex, Signals." Why would a burdensome and expensive reproductive procedure spawn itself so widely in nature? Why was sexual replication favoured to the detriment of the simple, neat, and relatively uncomplicated asexual reproduction? Moreover, why is that some species have kept both sexual and asexual replication? Sex must confer an advantage to sexual beings; otherwise, sexuality would have been eradicated from biological life. "Sex, Signals" not only answers such questions, but also makes the case that it is through sex that competition turns out to be a dominant trait in the overwhelming majority of living organisms. The cultural sphere cannot free itself from this. Sex breeds individuals in potential antagonism to one another. Any analysis of sexual beings interacting socially must consider the self-centred organism as its privileged unit. That is why cultural analyses focused only on the collective dimension of the group fail so consistently. Darwin's treatment of the puzzle of equal sex ratio in *The Descent of Men; and Selection in Relation to Sex* is quite close to the principles of game theory, albeit without using its vocabulary. "Sex, Signals" reflects on the consequences of the evolutionary arms race between parasites – which are asexual – and sexual organisms. Not culture, but the evolutionary pressure of mate selection establishes the urgency and the perseverance of communication among sexual beings.

The last part of "Sex, Signals" elaborates a signalling theory based on sexual selection. In direct opposition to Augustine's mentalist semiotics, it weighs the proliferation of signs in the natural world, taking as its starting point the immediate materiality of the signal. Peirce's classification of signs in relation to their objects is revaluated. Indices should be the focus of major considerations in a semiotic theory arising from the biology of sexual selection. It may be true that animals originally acquired signals to protect themselves from predators or to lure prey into deadly traps. Without a doubt, deception is evolutionarily advantageous. However, from an evolutionary standpoint one must also consider what happens to signals during the long expansion of time. It is inevitable that over time deception would be typically ineffective because of mutual decoding of the actors – senders and receivers – involved in the communication process. Combine the mutuality of communication with an evolutionary arms race and the result is a constant upgrading of the ability to detect deceptive signals. The evolution of truth is inevitable in both animal and human communication. Humans have a sophisticated critical faculty, and from it, the phenomenon of cultural change originated.

Strategies and players

Having established the inevitable priority of individual players, *Communication Games* then reflects on what marks the strategic decisions of players rationally equipped to take care of their self-interest. Although a theoretical abstraction not frequent in real life, the ground must be levelled: in this model both players have ideally the same capacities and the same goals, which means that the plans of action of each contender can only be reached taking into account the possible behaviour of the opponent. Does this mean that the cultural systems coordinating the players were unilaterally thrown away? That is not the case: both chapters in the section "Individual Games" of this book concede that the interaction of players occurs within cultural systems; but cultural systems – unlike the Augustinian God – are not powerful enforcers. In fact, social rules can be easily broken or twisted to serve individual interests. Culture provides the safety level for the interaction of players, and that is why in intercultural contacts social actors often experience a disturbing sense of personal discrepancy and social discomfort. So, if the cultural arena is a stage of antagonistic interest, its safety level is nothing more than the starting point of the conflict. It does not guarantee any result, which can favour any player.

After considering parlour games of the type tic-tac-toe and matching pennies, intentionally designed with the win/lose outcome in mind, *Communication Games* recognizes the existence of countless games in the natural and the human world where the result is of another kind: they are games that end with win/win, lose/lose situations. However, it should be remembered that cooperative games grow from an original competitive relationship. Because of the unflagging care of the players for their selfish interest, many games reach a point of equilibrium. In the history of game theory, this was called *the Nash equilibrium*. The supposition, initially demonstrated by John Nash half a century ago, is that in conflict of interests there is at least one point of stability from which no player has reason to depart.

While precarious and situational, the equilibrium is undeniable real. The 1:1 sex ratio in nature is a shining example of Nash equilibrium.

The notion of Nash equilibrium is extended to sharpen the definition of cultural systems. Cultural systems are balanced not because of the action of a controlling foundation: cultures are composed of countless points of equilibrium that are reached in respect to the self-interest of the players. Although the idea of equilibria is attractive to explain social stability, *Communication Games* indicates that Nash's conception can be criticized, showing situations where players choose strategies that, paradoxically, are not in their best interest. The objective is to fine-tune the early emphasis of game theorists in strategies to the detriment of players and signs and their quest for an optimal plan of action that may not be even feasible to ascertain.

Because of the disproportionate concern for abstract strategies, game theory neglected two important elements in the outcome of a conflict: the players and the information expressed in the signs circulating among them. It is from them that the solution of conflicts is reached. Players and signs are closely related: it is from their capacity to deliver a specific message that players are defining factors in games. The reception of signs is equally decisive in the constitution of a strategic plan. Knowing who their opponent is, players adjust their strategies to the potential capabilities of their contenders. If there is a dominant strategy, certainly measured by its capacity to invade other strategies, it is what is called *Chicken*. The players in conflict escalate and inflate their messages to a point that the opponent cannot reach. The loser then chickens out. In a competition, whoever is capable of delivering high-cost messages is the better player and the rightful winner of the game. The opponent recognizes that it is useless and against his most immediate interest to insist in the fray. Conflicts are solved through the display of messages, and equilibrium is then accomplished. However, one must bear in mind that inflated signs must preferably convey true information; otherwise, the player pays a high price for the deception. In cultural arenas, the display of wasteful, excessive, and inflated signs can solve conflicts before they even occur. Thus, it is not difficult to figure out why human languages are rule-oriented without ever doing away with the excesses of creation on the part of the speaker. Conventionalism cannot explain such a recurrent feature of the linguistic experience.

Communication Games concludes with a review of the Peloponnesian War to demonstrate that it is feasible to grasp cultural situations in a radically different way from conventionalist assumptions. Conventionalist dogmas are not necessary to interpret social facts. Without regret, conventionalism must be weeded out from semiotics of culture.[5]

Part 1
Canonical games

Chapter 1
Conflict

In the winter of 431-430 B.C.E., one year after the beginning of the Peloponnesian War, the Athenians gathered in the *agora* to participate in the public burial of those who had fallen in the conflict of Athens with Sparta and its allies. Despite being open to all living in the city, without discriminating against anyone, whether citizens or foreigners, the ritual of collectively burying the casualties in Athens itself, and not in the battlefields, is an exclusive trait of Athenian culture. Through the juxtaposition of blocs in Thucydides' narrative, the suggestion of a causal link between culture and conflict is forged in a conclusive way, and for the first time in classical literature. The explanation is seductive and has sounded, since then and frequently, adequate or right.

According to *The History of the Peloponnesian War*, the bones of the dead had come to Athens two days before the ceremony and were collected under a single tent where offerings were deposited. The procession followed from the civic heart of Athens in the direction of the Ceramicus in whose neighborhood was located the cemetery of the ones who fell in battles (*dêmosion sêma*); the crowd went through the way leading to Dypilon, the city's main gate, reaching the Academy (Gomme 1956b: 102), and then saw a line of graves along the road. At some point in the ceremony, an Athenian delivered a funeral oration (*epitahios*) to honor the dead heroes of the city. The Athenian council (*boulê*) had chosen the speaker, based on his intellectual worth and political reputation. At the end of Pericles' speech, and after the manifestations of grief on the part of the spouses and relatives of the dead, the Athenians carried on with their lives without realizing yet the destructive extent of a war that would last for twenty-six years more.

1.1. History, the speeches, and the funeral oration

The speech that Thucydides attributes to Pericles is memorable in its deviance from the conventional norms presiding over the *epitaphioi*. In the face of the endless discussion whether Pericles' funeral oration and all of the speeches in Thucydides' *History* were in fact pronounced under the compositional form that we have available today, the best admission is that anything written after the event cannot be an absolute record of each rhetorical detail. Thucydides himself, in the preface of his book (1. 22), admits writing the speeches without such intention. It would be impossible to remember the exact words of the speeches that he actually heard. Consider then how to rate the faithfulness of the ones that he collected through informants. The problem of perfect historical record is the same in each case, for memory dissolves the present, always reinventing episodes of the past.

If historical memory tends to be distorted, how could one achieve absolute fidelity, in a time when the great majority of testimony was oral, thus remade at each moment of recollection? Thucydides lived in an age when the written record was the exception, not the rule; and yet, this does not invalidate his narrative. For Gomme (1956a: 147), it is a false problem. Although the modern analyst can draw data from countless written records and hundreds of other historical documents, the historian's task is always to present a personal analysis, combining interpretation, narrative, and historical documents. A contemporary historian cannot be considered more faithful to the interpreted events than Thucydides can. In itself, the method should not be the problem.

On the other side of the spectrum, Simon Hornblower (1987) recognizes an insurmountable contradiction in the compositional method of *The History of the Peloponnesian War*. It would be impossible to conciliate the commitment to historical objectivity proclaimed by Thucydides himself with the product of his narrative so dependent on speeches to document facts, leading him into an inevitably subjective and strictly personal fabrication. Caught between the intention of historical objectivity and the personal choice of details, Thucydides declares that he transcribed in the speeches what seemed appropriate (*ta deonta*), an expression whose ambiguity was underscored by Hornblower. Hornblower noticed that literally Thucydides' words mean what is necessary. Perhaps the thing to do in this case is to keep always in mind a distinction between narrative and speeches. In fact, that was what Thucydides suggested. It is possible only to take full advantage of the speeches as modes of documentary evidence of what historical actors may have thought during the war (Rood 2002: 241). In both cases, Thucydides' text provides documents, and this does not mean that the modern reader has sunk into the anachronism of wishing to find the objectives of a modern historian in a classical author. Concerning the speeches, it would be more sensible to consider them in isolation, examining the appropriateness of each. Bosworth (2000: 2) claims that of all speeches in *The History of Peloponnesian War*, Pericles' funeral oration is the most faithful, for the *epitaphios* is firmly anchored in the historical context of the war.

The detour of Pericles' oration from the conventional repertoire ruling the *epitaphioi* indicates that Thucydides' reconstruction of the speech presented in the Ceramicus is not so distorted. In the memory of a chronicler, the impact of a novelty, small as it may be, would be greater than the obedience to conventions. But the dominant formula, in other words, the one that the historian would expect to find, was clearly transformed in Pericles' speech. The orthodoxy of conventions was then disfigured; and yet one can recognize a massive use of common places (*koinoi topoi*) in Thucydides' depiction of the funeral oration. Zolkowski (1981: 113) observes that of the thirty-nine common places identified in other funeral orations, thirty-one were present in Thucydides' *History*. That is no surprise: any orator delivering an *epitaphios* would tend to follow the rules that were extracted and imitated from the orations of the past. The conventions of the *epitaphioi* were inscribed in the minds of the Athenians. According to Loreaux (1986), the dismissal of the force of rules over the delivered speeches in Antiquity leads modern interpreters to the hasty conclusion that Lysias' and Demosthenes' speeches were not authentic for not having the habitual elegance of one and the biting spirit of the other. The model was sufficiently powerful to make extremely difficult the distinction of the speeches presented in the Ceramicus (Loreaux 1986: 10). Loreaux thus argues that the tendency was to have the personal mark of the orator abandoned because of the overwhelming impersonal force of the conventions.

The conventions of the funeral orations in Pericles' age determined the display of a tribute to the civic virtues of the dead with emphasis on the sublime and heroic sacrifice of the youth. The military past would be glorified. The audience would hear praise to the fearlessness of the ones who, although in small number, had defeated the massive Persian army in Marathon; and the speech would celebrate the citizens who, under the leadership of Themistocles, destroyed Xerxes' navy in the straight of Salamis. The speaker would not forget to underscore the altruism of the Athenians who deserted their houses so that the strategy of Salamis could succeed. Besides, the conventions established that the speech should have a section (*paramythia*), whose point was to console the relatives of the dead. The audience would hear the speech controlling their emotions, not enabling displays of grief and mourning. All would show resignation. The speech would also accentuate that the dead were an example for the living.

Nonetheless, Pericles' funeral oration takes a different turn. The speech avoids skillfully the themes of civic heroism and historical glories. In fact, what the *epitaphios* avoids is more significant than what it includes. Pericles had motives to keep away from references to the victories of the past. To talk about past triumphs would be an unnecessary political risk. The comparison would tarnish the present, for in the minds of many Athenians listening to what was said, the present troubles were to be blamed directly on Pericles.

1.2. Pericles' problems

A year before the delivery of the funeral oration, in the assembly (*ekklēsia*) that rejected Sparta's ultimatum, Pericles had argued that the appropriate strategy to defeat the enemy was to avoid confrontation on land. The Spartan hoplites were, in fact, more numerous and better soldiers than the Athenians, but Athens had in its favor greater naval superiority. To ensure Sparta's defeat, the Athenians should isolate themselves inside their walls, transforming the city into an island. In strategic terms, what Pericles proposed was more than reasonable, for up until the Peloponnesian War, the entrenched ones would have the upper hand in a conflict. Siege tactics were notoriously ineffective. At the time, no systematic knowledge about how to take a walled city by assault existed. Traditionally, the Greeks always fought in open spaces. The goal of an infantry battle was to make the enemy lose space and then disband because of its incapacity to resist the frontal attack of the victorious phalanx. In the case of besieged cities, the objective was to force rendition because of a lack of supplies. Besieging meant a long and costly struggle, the opposite of traditional Greek warfare that would generally not last long. A typical war was decided in a single battle and in one day (Kagan 2003: 1). Without catapults or other techniques of assault, it was difficult to invade a well-protected city like Athens. The Spartans would be frustrated waiting for a fast and immediate decision. Pericles' strategy was not surprising: Greek generals would rather win a battle without besieging (Ober 1985a: 45).

The price of Pericles' plan was certainly high. The Athenians would have to desert their properties outside the city walls; but leaving the city during warfare was nothing new. It had been Themistocles' strategy against Xerxes. Themistocles persuaded the Athenians to evacuate the city, forcing the Persians to face them in the naval battle of Salamis. It worked then; it had everything to be successful now. Although distant from the sea, neither the feeding and the supplying of the city nor the deploying of troops would be problems, for the robust walls of Athens extended to the port of Piraeus, from where the Athenians could easily displace themselves to fight or to import foreign products. The Athenian military, commercial, and naval power,[1] consolidated after its victory in Salamis, allowed Athens not to depend on the harvest of its own agricultural districts. It was a new kind of power in the Hellenic world (Finley 1963: 131). For this reason, in the funeral oration, Pericles would remind his audience that a city of Athens' greatness attracted products from all around that were then added to what was home-grown. The contribution of the agricultural districts to the welfare of the city could be dispensed shortly. Athens had the economic resources that were lacking in Sparta for a prolonged war. The strategy would require patience and self-control: eventually, without the Athenian monetary reserves, the enemy would capitulate.

But time went by and the balance of force between the cities in conflict brought with it an impasse that was undoing the morale of the Athenians. The Spartans had managed to camp at approximately ten kilometers north of Athens at the edge of the most populous agricultural district in Attica, the Acharnae. Rage and frustration grew among those protected by the walls of the city, especially the young Acharneans. Acharnea provided, according to Thucydides, a military force of 3,000 hoplites, roughly a seventh of all Athenians troops in Attica (Meier 1998: 469). The Acharneans then demanded action. Thucydides (2. 20) claims that the Spartan king, Archidamus II, had placed the Peloponnesian troops there on purpose, hoping that their presence would put pressure on the assembly, leading the Athenians to accept an infantry combat, which would be to Sparta's advantage. In the speech evaluating the Spartans' ultimatum, Pericles tried to prepare the spirit of those whom he imagined would revolt against the sight of their fields either burning or occupied by the enemy. To do that, Pericles reminded them that trees grow fast, but human beings are not easily replaced (Plutarch, *Per.* 33). Perhaps the adoption of Pericles' policy, so dependent on the charisma and the prestige of the one whom Thucydides (1. 39) said to be the leader of Athens and its most powerful citizen in both action and debate, demanded more than a metaphorical argument.

The fury of the Athenians was directed not only toward the Spartans, but also against Pericles. He was constantly singled out as the one who caused the ravage of the Athenian fields (Kagan 1991: 53). Probably to calm the Athenians, but perhaps mainly to avoid an assembly meeting that could topple his strategy, Pericles mitigated his policy of radical isolation. He ordered the Athenian soldiers to face the Peloponnesians. Or, as Josiah Ober (1985b: 181) suggested: in his argument to persuade the Acharneans to abandon their houses, Pericles may have vowed to protect their properties through the systematic use of the Athenian cavalry. Thucydides does not address this puzzle in *The History of the Peloponnesian War*. How was the peasant population, aware that a common tactic in Greek war was to burn and ravage the plantations of the enemy, persuaded to leave their fields? The historical precedent of Themistocles' successful strategy during the Persian wars cannot be a sufficient explanation. The argument aimed at the farmers had to be different. Pericles was asking for the sacrifice of some Athenians, not all of them. The farmers would have a natural resistance to the idea. Why should they give up everything, when the people from the city were deprived of nothing? The farmer's opposition could be diminished if Pericles promised them that the cavalry would protect their fields. The farmers knew that the horsemen would never defeat the Spartans but could well save their properties from greater damage, or, what was more likely, as Hanson demonstrates (1989 and 1998), the cavalry was deployed to protect the ancestral homes of the Athenians against the violation of their enemies. Because of the natural resilience of vines and olive trees, preserving the farms was less a strategy concerned with the economic welfare of the *polis* than the preservation of the Athenian sense of integrity.

The use of the cavalry to defend the fields was a feasible military plan. For an infantry to lay waste to an enemy's plantation, its soldiers would have to disperse. The hoplites would then scatter, undoing the rules of infantry alignment. Hoplite fighting followed a standard layout: the phalanxes were usually eight ranks deep, each soldier holding a shield (*hopla*) and depending for his safety on the one by his side sustaining his position (Cartledge 2003: 67). Without the bloc formation, each hoplite was more vulnerable to the sudden attack of Athenian horsemen (Spence 1990: 102). Cavalry incursions would not inflict huge losses on the hoplites laying waste to the fields. At best, one horseman could only kill one soldier at a time; that would be an ineffective strategy if the point were simply to fight the Spartan phalanx. The real purpose of the horsemen's attack could well have been just to disrupt the destruction of the fields and thus to protect them, albeit momentarily. Not much else could be obtained. No record of a definite victory of the Athenians over the Spartans in a situation such as this exists. Yet, in a war of constant attrition, Pericles' defense strategy based on cavalry charges had its own merits and was consistently employed during the engagements. Thucydides (7. 27) notes that the action of the Athenian cavalry increased as the conflict developed; in 413 B.C.E., the horses patrolled the fields daily to the point of exhausting many of them.

If the promise to protect the fields was made, it was not always successful. Thucydides (2. 21) tells of a cavalry battle with the Boeotians in which the apparent victory of the Athenians was turned into a surprising defeat. The enemy then placed a trophy not very far from Athens' Long Walls. It is reasonable to assume that the event was deemed a bad omen. Once more, Pericles was responsible for all this. He rejected the proposals of the Spartans. He triggered the war. In Athens, everyone seemed to have motives for resentment and anxiety. The urban population saw the streets swollen with peasants. The peasants would compare inevitably the luxury of the city with the hard routine of the fields. For the ones who had moved from the agricultural districts to the Athens, the abandonment and the destruction of their properties was not the only grievance: the farmers had to live as a favor with relatives or crowd themselves in precarious, unhealthy lodgings put hastily in place (Thucydides 2. 56).

Pericles had to sweeten the bitter policies that he proposed for the city, now at the brink of rupture. He knew that he was the constant mark of oligarch political adversaries, led by Thucydides, son of Melesias, probably from the same family as the author of *The History of the Peloponnesian War*. The Athenian political panorama was quite complex and should not be reduced to the oscillation between oligarchic and democratic extremes. Pericles' strategy of isolation inside the Long Walls brought with it a mutation in the process of defining public policies. Other interests emerged with considerable and inconstant force, a direct effect of the demographic storm that swept the political landscape in Athens during the year of the war. A skilled politician who had survived for so long at the crest of the inevitable instability of Athenian direct and non-representative democracy would need an acute sense for everything that could swell into a wave of criticism. It was the same with Pericles. He avoided public appearances, except when necessary, for he knew that he had similar facial features to Pisistratus, the tyrant whose dynasty spread violence and revolt in Athens. Pericles was aware that it was easy to accuse him of being a despot posing as a democrat: he was rich and noble. Perhaps, because of that, Plato had charged him with a lack of authenticity; as in the dialogue *Menexenus* (236 b), when Socrates claims that Aspasia, a concubine and a foreigner from Miletus, was the real author of one of Pericles' funeral orations. In the same dialogue, Plato suggested scornfully that the Athenian government rests over a fundamental falsity: it looks like a democracy, but it has always been an aristocratic regime, a ruling of the best supported by many (*Menexenus* 238 b). Furthermore, in *Gorgias* (515 c), Socrates' condemnation is much stronger: because of Pericles, the Athenians have become lazy cowards, loud-mouthed and envious.

For many Athenians, Pericles had other deficiencies. Cautious in conducting war campaigns, he was not a general who could display a spectacular military victory. He was not a pleasant person and had associations with suspicious individuals, polemic intellectuals, and agnostics such as Protagoras of Abdera to whom he gave the task of co-writing the constitution of Thurii, one of the Athenian colonies in the region of Lucania. He was a friend of Anaxagoras of Clazomenae (Kagan 1969: 142). Accused of impiety, both Protagoras and Anaxagoras were expelled from Athens. Pericles was under relentless attack. His plan of restoring Athens was severely criticized; it was said that he decked out the city like a garish prostitute (Plutarch, *Per.* 12. 4). He was accused of misusing public funds.

Pericles, however, always moved fast and with political competence, many times forecasting precisely from where and how criticism would come. Thus, even before the Peloponnesians invaded Attica, Pericles guessed that the king of Sparta, his friend Archidamus II,[2] perhaps would spare his properties from being laid waste as a token of friendship or else throw the Athenians against his authority. In an eloquent gesture, with a clear persuasive purpose and duly recorded by Thucydides (2. 13), Pericles transformed his land into public property. It was now difficult to oppose his suggestion of abandoning the agricultural districts: he was doing with his own possessions what he suggested to the Athenians. It was not an impossible sacrifice. The land donation was an answer to any future criticism.

1.3. What to praise

Facing a dangerously unstable political situation and having to meet the expectations of the crowd at the cemetery, Pericles delivered a speech mistakenly seen by later commentators as an abstraction, a description of the ideal democracy that Athens ought to personify.[3] In both cases, the speech would be deemed an invention, but there are some questions to be answered if one considers the *epitaphios* relatively unfaithful or as a well-intentioned construction meant to uplift the spirit of a new generation of Athenians hardened by the unfolding of a seemingly endless war and thus immersed in cynicism and disillusion. How is it possible to conciliate these assessments with the confessed goal of factual objectivity that inspired Thucydides (1. 21)? How could Thucydides distort or falsify an event as public as Pericles' funeral oration if some of survivors of 431 were still alive to refute what he presumably wrote in 401 B.C.E. (Zolkowski 1985: 192)?

It is inconceivable that a perceptive politician like Pericles would have made a public pronouncement without taking into account its consequences, and not having in mind a further motive or purpose. That is more so if one notes that the *epitaphios* avoids carefully what is disputable or polemic. Persuasion is always what defines Pericles' public actions. When asked about who was the better wrestler, Archidamus answered: every time I threw Pericles on the ground, he argued against it and persuaded the audience that I was the one thrown down (Plutarch, *Per.* 8).

The choice of contents displayed in the funeral oration takes into consideration the rules guiding the genre, but what Pericles effectively does is to present a specific answer for that moment in Athenian life. One issue made him uneasy: how to represent the transformation of individual loss into a sentiment of civic glory demanded by the conventions of the *epitaphioi* without personalizing the felling and thus evoking in the Athenian audience the hardships that they were going through? In Thucydides' account of Pericles' speech, each step in the oration makes what was expected of all *epitaphioi*: to minimize and preferably to deny the divisions within the city (Loreaux 1986: 331). In a general way, Pericles' funeral oration reestablishes order without resorting to hierarchy, suggesting the union of Athenians through the exaltation of equality.

Perhaps Pericles delivered the speech having in mind the lessons he drew from the encounters with Protagoras. Through the echoes of the malicious gossip of Xantippus, Pericles' own son, Plutarch (*Per.* 36) mentions endless discussions between the politician and the sophist. Pericles and Protagoras examined exhaustively, to the point of absurdity, the contradictory sides of a dispute. It was definitely not from the influence of sophist Gorgias' antilogical rhetorical style that Pericles structured his speeches to the Athenian assembly. Historically, that could never have been the case, for Gorgias had only visited Athens in 429 B.C.E., two years after Pericles' death. Yet Protagoras' influence over Athenian culture was ubiquitous. John H. Finley's definite historical analysis of Thucydides' rhetorical style established that antilogy and antithesis were the dominant stylistic traits in Athens (1967: 55–117), recognizable in sophists and orators such as Thrasimacus and Antiphon or dramatists like Sophocles and Euripedes. Over that whole period and thus over Pericles' himself, the shadow of Protagoras and his fondness for antilogies projected themselves.

Antilogies embody a wider conception of the logical structure of speech, precisely the cognitive posture that Protagoras disseminated. It is known that he taught his disciples to attack and defend, to accuse and excuse, to celebrate and condemn the same viewpoint (Romilly 1991: 76). Protagoras' sophistic teaching professed that every discourse branches out into another one that opposes it. He also proposed a solution for the impasse in the face of antilogical options, stating that the argument expressing the wider consensus always prevails. Unable to accept the existence of an absolute and irrefutable truth, Protagoras repudiated the possibility of anything transcending the frailty and the precariousness of what is strictly human. He put between brackets the possibility of knowing the divine order and saw all knowledge restricted to a universe of opinions where everything is equally valid and in which it is impossible to decide what is right or wrong, true or false. The ensuing conclusion is that truth depends exclusively on a majority consensus that confers force to a speech. That is how Plato (*Theaetetus*,

172 b–c) describes what Protagoras thought was the truth. Consensus grants truth-value to what is said. Before addressing the crowd in the cemetery, Pericles certainly tried to find an advantageous argumentative point from which he could answer his critics. He had to use a consensual and conventionalized force that would silence the objections of his critics.

The authors of ancient rhetorical manuals, such as Pseudo-Dyonisus and Menandrus and modern interpreters (Zolkowski 1985), have noted that Pericles' funeral oration discarded a section of the formulaic *epitaphioi*, which is the lament (*threnos*) for the dead. If one excludes the *proemium* (Thucydides 2. 36-42) and the epilogue (2. 46), which would present be in any speech, Pericles' oration is divided into two segments of disproportionate sizes. The larger one (Thucydides 2. 36-42) displays the praise (*epainos*) of those who fell in combat against the enemies of the Athenians, and the other (Thucyidides 2. 43-45), just a fourth of the largest segment, under the form of advice, is an exhortation (*protepticus*) to the living, whether the children of Athens, therefore the future of the city, or the brothers of the ones who died, or whose citizens who could soon confront death. The oration also delivers cold words of compassion to the parents and relatives of the dead. Zolkowski (1985: 57) shows comparatively that in all of the extant funeral orations from Antiquity, whether civic or individual, the lament (*threnos*) is absent in most of them, with the exception of Lysias' oration. Although the examples do not display the lament, the manuals recommend them. The reason that there is a discrepancy between the recipe and the illustrations may be because the lament was merely a conventional option, not a mandatory prescription in funeral orations. Pericles' *epitaphios* does not break any rules of rhetorical composition. Its option is to suppress the lament because of the personal and political interest of the speaker. Pericles' must have feared its negative effect over the audience. Several sentences sprinkled at the end of the speech can be seen as faint ripples of a lament: "as for those here who are brothers of the dead, I can see a hard struggle in front of you" (Thucydides 2.45). After promising that until their adulthood the sons of the deceased will be supported by the city, Pericles affirms: "This is the crown which she offers, both to the dead and to their children, for the ordeals which they have faced" (Thucydides 2. 46).

Pericles' oration begins with a negative granting the antilogical tone of the speech. It is as if Pericles is denying the relevance of what he is going to declare. Although the ceremony of funeral praise is a custom of the city, the glory of those being buried does not need speeches to qualify it. No words can express the greatness of the dead. Furthermore, not only the greatness of heroes can dispense with *epitaphioi*. Praise can be below the actions or even overvalue them. However, acts can only be honored with other acts (Hornblower 1991, v. 1: 296). In both cases, the audience would be divided by envy or incredulity; yet, despite his apparent refusal, Pericles obeys the tradition and meets the expectations of the ones gathering in the cemetery.

In Thucydides' narrative, the section devoted to the traditional praise of the dead refers briefly to the events of the past and ends up describing why Athens is splendorous. If the more remote ancestors deserved praise, the parents of the dead would have to receive greater praise. With their blood and toil, they created the empire that is the pride of all in that crowd. Avoiding the comparison of the present with the historical deeds of the past, Pericles begins to celebrate the actual state of the fatherland (*patrios*). The praise of the ones who fell in the strife with Sparta is changed into the cultural celebration of Athens.

1.4. In praise of Athenian culture

In Thucydides' rendition of Pericles' speech, Athens is a unique city. In contrast with Sparta, which borrowed its constitution from Creta, the Athenian governmental system is original. Athens serves as a model and a school for the Greeks (Thucydides 2.41). The city is a living lesson for anyone. Pericles proclaims that the magnificence of Athens derives from its democratic culture (*paideia*).

The funeral oration argues that the democratic order refuses to allocate the power of the city to the hands of a minority; instead, power is spread throughout the social body. Athenian equality manifests itself not only in mere legal disputes between individuals, but also in the establishment of those who fulfill a public function. Inside Athens, "what counts is not membership of a particular class, but the actual ability which the man possess" (Thucydides 2. 37). Being rich or poor is of no importance. All Athenians have the prestigious opportunity of serving their city, regardless of economic condition. Here, as in so much of what is said in the *epitaphios*, Pericles is not just praising democracy: there in the cemetery, everyone knows that he is also defending his policy of paying a daily wage for the citizens who dedicate themselves to public service. It is Pericles who in another political meeting proposed and saw that the assembly would approve of paying two and a half obols per diem to the Athenians participating as jurors in the tribunals: the payment represented half of a daily wage of a worker. Despite being criticized by many Athenians who saw this as a blatant bribe, the payment for public service is a practical and innovative measure: it allows the poor to get directly involved in the administration of the city.

In Athens there is no privilege granted prior to the assessment of the individual's merit. Civic virtue (*areté*) defines each Athenian. The *areté* is individually conquered in the confrontation (*agón*) with and the comparison to other citizens.[4] It is a measure of individual excellence and it is available to anyone who deserves it. In the ancient Hellenic world, there is no other city embracing this concept of citizenship. Thus, Pericles alleges that the civic value of an individual in Athens comes from the culture ruling and shaping him.

Culture has an unbridled force. It is easy to see it and taste it; it is constantly present in "all kinds of recreations of our spirit" (Thucydides 2. 38). For the Athenians, Pericles goes on to proclaim, personal wealth is just a useful resource, never a reason for self-glorification or a tool for public humiliation. In Athens, it is not a shame to be poor. There is only shame in the individual's incapacity of improving his personal destitution (Thucydides 2. 40). Each Athenian is the master of his own destiny.

Democratic culture also answers for the military prowess of the city. Pericles assures that the Athenians are better in war than their enemies. The courage of the Spartans is the result of an education based on scarcity and shored up by authoritarian discipline. Spartan militarism resides in the fear of the smallest weakness. Nothing like that happens in Athens. The Athenians make themselves stronger in the exercise of their daily life, forging a natural courage because it is not induced from above. Contrary to Sparta, Athens is open to all foreigners: it has no fear of anyone knowing its secrets.

The city is diverse because of its acceptance of multiplicity. Athens' versatility reveals itself not only in its democratic mode of government through the free discussion of public policies, but also in the way others are treated when guests of the city. As a result, the military force of Athens is split between naval and hoplite power. Any Athenian defeat in the hands of its enemies cannot be considered a victory over the whole of Athens. So far, the Spartans only sampled a fraction of what the city can do in war. This displacement in Pericles' oration is subtle and effective: it justifies previous and future defeats in the struggle with Sparta and, at the same time, affirms the Athenian superiority despite what had been happening up until then. The antithesis is clear: even in the case of defeat, Athens is victorious.

Indeed, the citizen-soldiers buried in this cemetery personify the idea that it is righteous to die for a city like Athens. Athens can only be deemed splendorous because its citizens are willing to die for it. Collective and ample discussion precedes any offensive deployments, defensive strategies, or sacrifice for the city. Athenians accept the compatibility of words and deeds (Thucydides 2. 40). What is the point of acting without understanding the effects of the strategies? Through democratic argumentation, the risk involved in the implementation of specific public policies is carefully evaluated. Athenian bravery has nothing to do with impetuosity and ignorance.

Pericles then progresses in his speech, blending the courage of the Athenian citizen-soldiers with the city's splendor (Gomme 1956b: 135). The funeral oration declares that the Athenian standards are undoubtedly high; for the sake of Athens, everything is demanded, even the supreme personal sacrifice. But this could only be a happy death, for freedom is liberty, and liberty, courage (Thucydides 2. 43). Loss and sacrifice must be understood in another way: they are acts of fortune. Here, in *The History of the Peloponnesian War*, one sees another example of the argumentation through contraries, using antilogies so much cherished by both Pericles and Thucydides.

Athenian democratic culture is the reverse of its enemies' way of life. The incompatibility between cities is complete: Thucydides displays Athens and Sparta as absolute antilogies. During the Spartan assembly preceding the delivery of the ultimatum to Athens, Archidamus II does more than evaluate what may happen. In Thucydides' narrative, the experienced and moderate Spartan king forecasts a hard and difficult campaign against a rich city, extremely well equipped, with the largest population in the Hellenic world and the most powerful navy with the best sailors in Greece (Thucydides 1. 80). After considering the objective conditions of the coming conflict and recommending the postponement of the hostilities, Archidamus evaluates Spartan and Athenian cultures. Athenian courage is frail when compared to Spartan valor. In Sparta, to be courageous is to be obedient and submissive. For that reason, the Spartans do not engage in the convoluted debates that enchant the Athenians. To ponder or to argue about courage is of no importance. What matters on a battlefield is simply to act as a brave and courageous citizen. Discipline makes an effective soldier of every Spartan. Laws, habits, and orders are not questioned but obeyed (Thucydides 1. 84). Sparta would rather keep itself in a pristine state, cultivating its traditions and authorities.

Later, at the brink of outright conflict, in another speech, Archidamus compares once more Athens with Sparta and foresees the strategy of isolation inside the Long Walls that will be eventually proposed by Pericles.

Archidamus supposes that, used to pleasure and enjoyment, the Athenians will not hold on to their policy of isolation after seeing their properties burnt and destroyed; they cannot bear losses. The staunch individualism of the Athenians will prevail, and they will certainly try to defend selfishly their personal interests in detriment to what is best for the city. Spartan strength, however, springs from the fact that the city not only has a large number of hoplites, but its army is composed of resolute citizens who are disciplined and willing to sacrifice themselves. The rigid and closed culture of Sparta allows many to act in harmony as though they were one (Thucydides 2. 11).

Another important antilogical juxtaposition in Pericles' funeral oration and directly related to the comparison that Archidamus makes of the Spartan and Athenian culture is the conjunction of liberty and discipline. Athens is free in its submission to discipline. To argue in favor of this idea, Pericles indicates that Athenian tolerance is not a drawback; it comes from obedience to its laws. In the city, no room for unfettered, savage, and anarchic individualism exists. Democratic freedom requires a special kind of self-control and compliance to the several contracts and social agreements, which complement each other, such as the right to criticize public policies and the duty to honor the decision of the preponderant vote in assemblies. In that way, and quite different from Sparta, Athens is transformed into a whole which is much more than the simple addition of individual minds.

Nonetheless, according to what Pericles professes in the funeral oration, it is incorrect to suppose that he is describing what is typical of political life in Athens. Being a regime without representation and fixed mandates, Athenian direct democracy is constantly at the mercy of demagogic thrusts, frequently led by the selfish interest of whoever is capable of captivating the assembly. The whole is done and undone at the sway of circumstances.

There is a paradox at the core of the democratic experience. Democracy depends on communal decisions, but the imprint of its direction comes from individuals who stand above the crowd. The political leader in a direct democracy should have a clear and steadfast vision of what is to be done as well as charisma, persuasive rhetorical resources, social prestige, and reputation with the assembly. As the narrative of Thucydides demonstrates, after Pericles' death, Athens will be steered into tantalizing whirlwinds. The end of it all will be a plunge into chaos and defeat.

If the funeral oration is neither a description nor an idealization of Athens, the best way to understand it is to consider it a political instrument of that moment. It is feasible to see the praise of Athenian culture as a celebration of its imperial power (Romilly 1963: 131). Up-to-date vocabulary would attribute to cultural traits what Jacqueline de Romilly called the spirituality of Athens. With this in mind and also considering the speech an answer to the present troubles of the city, Cogan (1981: 41) recognizes that the oration is appropriate not only for Pericles' intentions, but also for appeasing the delicate situation of Athens in the first year of the war. Pericles creates an image of Athens free from material possessions. Although it may have caused losses and even the destruction of the Athenian agricultural districts, the strategy of isolating the city inside its walls is more than merely appropriate; it is justified and necessary. It is perfectly in keeping with the spirit of Athens. Here one can find traces of Anaxagoras' influence over Pericles. It is noteworthy that in the funeral oration, one does not find the slightest reference to the Parthenon and other material forms of greatness that were widely known products of Pericles' public commitments. Athenian greatness is of another kind. One ought to call it mental, the expression of intelligence, of the *Nous*, claimed by Anaxagoras to be the supreme force, present everywhere, and ruling over matter.

The cities are their culture. This insight is valid for both Athens and Sparta. Thucydides does more than assert the antagonism of Athenian and Spartan culture. For him it is obvious that Athens is superior to its enemy. The casual link between cultures and group conflicts emerges. One will not find, it is true, this generalization literally stated in *The History of the Peloponnesian War*; yet it is a perfectly coherent inference, although more derived than directly declared by Thucydides. In any case, reading Thucydides, one clearly recognizes that the difference in temperament between Athenians and Spartans has a deep influence in the course of the war (Westlake 1968: 122). One can ask, what caused such a difference between Athenians and Spartans? The answer found in *The History of the Peloponnesian War* is the culture of each city. This inference matches Thucydides' style of explanation. For what other reason would Jacqueline de Romilly (1967: 11) point out that the establishment of a sequence in the historical narrative of Thucydides, even sometimes a turn of phrase connecting described facts, insinuates a possible interpretation to be figured by the readers? Under the guidance of Thucydides, it is tempting to conclude that cultural differences pit social groups against one another.

1.5. The city in crisis

Thucydides' narrative offers no clue to the reactions of the Athenians to Pericles' speech. The scene of the funeral oration ends, and *The History of the Peloponnesian War* jumps briskly to the next summer. Two-thirds of the Peloponnesian army had invaded Attica under the command of Archidamus and began to destroy the fields around the city. Thucydides' silence about the reactions of the Athenians to Pericles' funeral oration is a possible indication that they were resigned to accepting his advice. Before the delivery of the *epitaphios* in the cemetery, Pericles had suggested that the Athenians should destroy their properties to convince the Spartans that they were willing to give up their belongings. Now, the Athenians would have to bear with fortitude the view of a horizon of flames along the walls of the city. The farmers' houses, made with tiles and bricks and sustained by a wooden frame certainly would collapse when burnt (Hanson 1998: 72). Away from their fields, the Athenian farmers would see coils of smoke coming from their homesteads and then easily guess what was happening to them. That was not everything; the worse was yet to come.

After mentioning the Peloponnesian invasion of Attica, Thucydides tells how, two days later, the plague broke out in Athens. The initial signs of the pestilence came from Lemnos, an island of the northern Aegean Sea, which was an Athenian colony. At first, the symptoms of the plague were moderate, but they became more and more virulent. The Athenian doctors did not know what to do with the unfamiliar disease. The plague spread thorough contagion. The Athenians thought that the Peloponnesians had poisoned their water supplies. Thucydides (2. 48) describes the symptoms of the plague that had infected him also. The eyes were inflamed. The mouth smelled bad and oozed out with blood. The sick threw up in violent spasms. Fever burnt incessantly, and the skin turned red in patches. Pustules and sores erupted all over the body. The infected were not content with drinking water constantly; they dived into the city wells to lessen the affliction consuming them. Domestic animals such as dogs died, and no birds of prey circled the Athenian skies: "Though there were many dead bodies lying around unburied, the birds and animals that eat human flesh either did not come near them, or, if they did taste the flesh, died of it afterward" (Thucydides 2. 50). Thucydides remarks that the catastrophe had devastating social effects. Incapable of foreseeing what lay ahead, the Athenians increasingly disdained the laws and the customs of the city. Athens's cultural fabric, precisely that which, according to Pericles, bestowed force, discipline and social cohesion, was damaged. Athenian society and culture were in agony. Contempt, despair, and neglect of religious duties were seen frequently in the city's daily routine.

From a modern historical viewpoint, the way to deal with Athenian plague has been to ask what disease wreaked havoc on the city during that summer of 430 B.C.E. Did Thucydides make up the symptoms from his reading of Hippocratic medical manuals? Or had he portrayed, with minutia and precision, a real pestilence? And if so, what was it? It is indeed proper to have these questions together. Thucydides did use the medical jargon of the times and was the first Greek author who refused to take mythical sources, or any other accepted fiction, as roots of historical interpretation. He always evaluated the past with standards of factual objectivity to a point that one can deem *The History of the Peloponnesian War* a force that shapes up "a new definition, a new theory of truth" (Williams 2002: 163).

Thus, considering the publications up until the end of the 1950s, Gomme (1956b: 145-162) reviews the main assesments of Thucydides' description of the plague that regard it a true and real disease. Much later, Poole and Holladay (1979: 282) try to produce a radical reevaluation of what *The History of the Peloponnesian War* had said about this medical issue. After observing accurately that Thucydides seems impatient with the current conflicting ideas about the pestilence in Athens, Gomme appraises the modern identifications of the disease. Was it ergotism, measles, or typhoid fever? Gomme concludes his review unsure about the cause of the plague. Poole and Holladay doubt if it is possible to identify the symptoms of an infection described two thousand years ago with the parameters of today's illnesses. Not only do microbial diseases evolve and suffer radical mutations, but medical knowledge and its descriptive frame also change dramatically. Besides considering the hypothesis that Gomme evaluated, and agreeing with him, Poole and Hollyday discard smallpox, bubonic plague, scarlet fever, and a combination of several diseases. They argue that the Athenian plague could very well be extinct, or it could have suffered such mutation that one cannot match it perfectly with any modern infection. Much more important than recognizing the cause of the pestilence that devastated Athens is admitting and admiring Thucydides' precise and objective observation. His description singled out two aspects of the plague that would have escaped the regular medical literature of the Ancient world: contagion as a mechanism of propagating a disease[5] and the phenomenon of acquired immunity, for *The History of the Peloponnesian War* records that the infected, if they could survive the plague, would never become sick with the same malady again. Is Thucydides' narrative no more than a record of historical events? If so, it may be appropriate just to inquire if the historian accurately described facts. However, one should not forget that *The History of the Peloponnesian War* is also a literary construction, a text with meaningful compositional order. Because of this, the chapter on the Athenian plague is more than a mere description. The very placement of the chapter in Thucydides' narrative predicts the unraveling of Athenian democracy; the plague inaugurates a state of disregard for the laws and cultural conventions of the city never seen before. The narrative of the plague is thus in perfect agreement with Thucydides' insinuation in other parts of his *History*, giving to cultural factors a dominant role in the Peloponnesian conflict. For the first time in Athenian life, refrain in displaying riches is forgotten. Wealth is exhibited recklessly. Lavish expenditure becomes a common indulgence.

The plague infects the confidence and hope illuminating Pericles' funeral oration. It is striking that one of the identified victims of the plague is Pericles himself. The plague consumed him slowly. He became an altogether different person. The Athenians would then see their leader, in the past so devoted to praising the power of rational thought, carrying a magical tasliman around his neck (Plutarch, *Per.* 38). Historians have suspected that Pericles' death had nothing to do with the plague because the disease decimated the infected much more rapidly than Pericles' agony lasted (Meier 1998: 472). Thucydides' (2. 60) narrative only mentions that Pericles died two and a half years after the beginning of the war and that his death was a irreparable loss for Athens during the rest of the war.

With the plague begins another phase in Athenian history, marked by bitter regret. Pericles' prestige fluctuates. More than ever, many citizens blame their misfortune on him. They cannot perceive how the strategy of isolation inside the Long Walls was more than a mere strategy to win the war. Regardless of its unintended consequences, the isolation was in fact a persuasive device whose purpose was to convince the Spartans of the need to reach an agreement in which none of the two great powers of the time would try to subjugate the other (Kagan 1991: 220). For each Athenian citizen, however, what mattered most was the immediate pressure of present troubles. What was the sense in trying to reach a future goal, a distant peace? Why should they try to persuade the Spartans of the futility of wars when the present was so distressful?

The plague hit hard on the military resources of Athens, killing 4400 hoplites, 300 horsemen, and an incalculable number of citizens from the lower classes, wiping out probably one-third of the entire population of the city (Kagan 1974: 71). In the end, the plague neutralizes one of the military goals of the isolation, which is to preserve to the maximum the human resources that managed the Athenian ships. It is unrealistic to expect that, in the midst of such extensive losses, the Athenians came to realize how brilliant and innovative Pericles' strategy was. Like so many times before, Pericles walked a tightrope with his policy. When the most important political leader in Athens presented a rational war plan, implementation depended upon the agreement of the assembly, whose choice of public policies was seldom completely based on reason.

1.6. The answer before dying

To make matters worse, the strategy of isolation inside the Long Walls stood in direct disagreement with the military conventions of the Greeks. The price the Athenians would have to pay to hold their ground and abide by Pericles' plan was more than the desolation of their agricultural districts. To refuse combat was considered a plain admission of cowardice: one would suffer the loss of individual honor. In a context of dejection and discouragement, Pericles' last speech in Thucydides' *History* aims at not only stimulating the crowds, but also inventing the necessity of mustering collective courage, a daunting task in a world where bravery distinguished and discriminated individuals, granting them social glory and civic virtue.

The idea of public courage is at odds with the notion of bravery celebrated in the Homeric poems, which Thucydides (introduction. 10) sees as fantasies, or at best, exaggerations. In the first chapter of *The Iliad*, after an explosion of rage, Achilles insults Agamemnon, the commander-in-chief of the Greek expedition against Troy. Achilles neither respects nor obeys the commander of his army. Achilles and Agamemnon are individuals of a relatively equal status. Any distinction between them would have to come from the greatness of deeds, not from a hierarchical position such as commanding the army. At this moment, hierarchical distinction is pointless. For the Greeks, nothing is more important than the value of the individual, and Achilles knows how much his deeds are generally admired. In short, running and overcoming risks define the hero. It is through undergoing what few could go through that one establishes one's worth. In Greek wars, the ones who defy death face-to-face are the ones to be celebrated heroically. The common assumption is that the archer, the javelin thrower, the slinger, anyone who fights from the distance should be despised. To suffer wounds in this way is the result of cowardly blows (Hanson 1989: 16). Thus, the Cretan captain Idomeneus (*Iliad* 13. 306–313), shows contempt for spears, denying that it is his intention to fight at a distance, out of the enemy's range; a hero ought to plunder with his full armor polished bright. Whoever participates indirectly in battles, throwing from the distance lances and arrows, however much he contributed to the final victory, is not to be honored.

In the midst of the confrontation between Achilles and Agamemnon, Nestor, the king of Pilos and a member of the Greek army, tries to intercede and put an end to the squabble. Nestor's call to reason does not evoke the urge to respect or obey the commander of the Greek expedition. Nestor asks both Achilles and Agamemnon to leave their resentment behind. This emphasizes the impression of equality between the two men. Neither Achilles nor Agamemnon listens to Nestor's appeals. Achilles leaves the place even more enraged. He asks his mother, the goddess Thetis, to intercede with Zeus and inflict defeats on the Greeks (*Iliad* 1, 590-609). Achilles prays with fervor that his fellow-soldiers be annihilated. In the narrative of the poem, there is no hint that Achilles' resentful curse on his own people diminishes his heroic stature or that it is a reproachable act.

During the Peloponnesian War, another conception of heroism is offered to the Athenian citizen-soldier. Not the individual, but the city, the group, indeed the social whole, becomes the potential place for the manifestation of the *areté*. Thucydides indicates that, for Pericles, the heroic model is not anymore the one inspired by Homeric narratives. The funeral oration proclaims that the *polis* does not need the praise of Homer (Thucydides 2. 41). This is not, as has been suggested (Brunt 1993: 165), a merely arbitrary insertion in the *epitaphios*. Even if it were an entire attribution on the part of Thucydides, the phrase represents an alternative conception of military valor.

With Pericles, military planning begins to take the place of the ritualized strife that used to define traditional Greek battles. To grasp the social sense of such a ritualized confrontation, one must take into account that although all in the group follow the formal rules, here what truly matters is not the concerted unity of the whole. In reference to this unifying pattern, it is possible to single out the best fighter, the most prized individual soldier, in fact the hero. Whoever wins or shines in a confrontation, or in any competition, abiding to what the norms allow, without cheating or bluffing, is the rightful winner. He is an individual of true excellence. His social prize is deserved; the honor is legitimate.

The Greek army is centered on the individual. The minimal principles of cooperation that bear some collective traits and that are easily seen in hoplite fighting surface from the individual acts of each soldier. Although it is of minor importance to lose the protection of the chest and the helmet during the fray, it is an unambiguous shame to drop one's shield during a battle. The shields create a protective bloc for the whole phalanx; the other parts of the full armor only guard the individual (Spence 1990: 98). The breastplate and the helmet cannot be so crucial for fighting.

Although group formation protects the otherwise lone and vulnerable soldier, it is from the conjunction of individuals that the phalanx maintains its shape.[6] In this context, one must understand the symbolic value granted to the shield in Greek body armor. In *The lliad*, after hearing that the Trojans had desecrated the corpse of his beloved friend Patroclus, Achilles decides to return to fighting. His weapons are unavailable. He had given them to Patroclus. All his armor is now with his dead friend's body held by the Trojans; he needs a set of weapons. At his request, Achilles' mother commissions a new armor from Hephaestos, the greatest craftsman, the god of fire, the smith of the gods. Hephaestos forges magnificent pieces, a full armor: a shield, a helmet, and grieves to protect Achilles' legs (*Iliad* 18, 525-527). It is rather telling that one of the pieces, the shield, merits minute and special description in detriment to the rest of the armor. Homer's verses indicate that Achilles' shield is a dazzling cosmos, a visible indication of the excellence of the hero who, because of his own choice, will not live for long.

The shield is the main piece of hoplite armor. It is heavy and cumbersome, but it is also an identifying mark of belligerent heroism, clearly distinct from the part of armor that defines the one whom Achilles will slaughter: the Trojan Hector. Hector's helmet is a constant reference of his presence, not the shield. It had to be that way. Hector is a brave and brilliant warrior, but in several parts of *The Iliad*, he is depicted as father, son, and husband. He is more than a fighter, so very different from Achilles, who is just a dog of war, blind and tormented by the death of Patroclus. Achilles' status is exclusively combative. The shield is his most appropriate representation.

When Achilles moves to kill Hector and finally to revenge Patroclus' death, Hector backs down, and thinks of going to the dueling field without weapons, without his shield. Hector was ready to offer the return of Helen, even the treasures that his brother Paris, plundered from the house of Atreus (*Iliad* 22, 132–150). Hector muses: the war must end. Fear overcomes him, and sure of Achilles' obsession, Hector runs from the fight. He goes around the Trojan walls until Achilles finds him and cuts his throat. During all the description of Hector dying, *The Iliad* mentions repetitively his flaming helmet.

The importance given to the shield is more than a mere residue of ancient traditions, renewed in the narrative of Homer. As a symbol of military prowess, the shield remains revered along the course of Greek history. In his commentary about the symbolic elements in Hellenic landed battles, Connor (1988: 25–26) points out that, after the pitched confrontations, the hoplites would return home exhibiting their shields as emblems of military glory. A campaign would only end with the dedication of the shields in the altars of the *polis*. Centuries after Homer, in his plays *The Birds*, *The Clouds*, and *The Wasps*, Aristophanes pokes fun at Cleonymus, his political enemy, calling him a snitcher, a glutton, a coward who threw away his shield in the battle of Delion when the Boeotians defeated the Athenians. Staged in 422 B.C.E., *The Wasps* begins with Xanthis, a slave, describing, as a bad omen, a dream in which Cleonymus, once again, throws his shield away. In his *On Rhetoric* (2, 1383b, 22, 3), Aristotle exemplifies shame as "throwing away a shield, or fleeing in battle, for these come from cowardice."

With the exception of the Spartans, a Greek battalion had little or no training and drilling. No collective rehearsals before the fight occurred. The individual citizen bought his own hoplite armor, without help from the public coffers. The citizens with economic means were the only ones to fight in a phalanx. The expenditure in buying armor was considerable: the whole armor cost the equivalent of several monthly earnings of a moderately skilled craftsman (Connor 1988: 10). The praise and the glorification of the hoplites were more than an objective estimation of military acts. On the one hand, it is the residue of a conception of heroism that gives special weight to the individual hero; but, on the other hand, it is social prejudice, an idealization of the upper classes and the wealthy citizens. This preference downplays the rowers who came from lower social rungs, and who, in fact, constituted the new basis of political and military power in Athens.

The Athenian growing dependence on naval power helped change that picture. Public expenses became increasingly necessary for military engagements. The city-state, the social whole, more than the individual soldier, was transformed into a decisive element in the conduction of the conflict. For this reason, Pericles emphasizes the urge of having a collective plan for the war. His idea is that a sound strategy plus public funds will ensure the military success of Athens (Thucydides 2. 13). Only decades later, Athens will provide training and equipment for the citizen-soldiers. In *The Athenian Constitution* (XLII, 4), written between 328 and 325 B.C.E., Aristotle mentions that the Athenian cadets received "a shield and a spear from the state."

Planning and strategies did not come into existence during the Peloponnesian War (Krentz 1997), Thucydides' *History* does record several ambushes, strategies, and sneaky night attacks. However, one should always be careful when talking of planning during the Greek wars. Indeed, the method of combat follows another pattern, quite different from that of modern warfare. A Greek general would rather invoke the intervention of the gods or repeat with variations what had been done in the past. In this setting, strategic innovations are hard to grasp.

The battles between Greeks are not, as some may hastily think, a straightforward ritualized conflict with rules previously established and followed by all. Nonetheless, it is tempting to suggest that the fight obeyed rigid rules of behavior. In this case, the strife would stick to a neat sequence: after the challenge, the enemies would meet in a designated battlefield where the conflict would unfold, fiercely and savagely, ending with the victorious placing a trophy in the field. Then there would be a truce so that the defeated army could collect the remains of their dead.

In reality, the conflict is a squalid mess. One army moves in the direction of the enemy, both aligned in columns. The hoplites of each army know that it is necessary to run against the mass of soldiers in front of them; otherwise, it is impossible to penetrate with their spears the iron, the bronze, and the wood of the enemies' armors. At the same time, running helps each individual soldier to avoid the projectiles thrown from a distance. Then, the concerted alignment of the groups begins to undo itself. The war chants and cries, the music of the Spartan flutes are attempts to coordinate the movements of each hoplite. It is difficult to hear the instructions of the commanders. The armies rely on individual orders passing from one to another in the battlefield; this only leads to the disruption of the alignment. The phalanx is always on the brink of turning into a shapeless mass. How could it be any other way if each running soldier, some young, some old, has diverse physical conditions? Yet, to guarantee the safety of the phalanx, the ordered alignment ought to be kept intact. If not, the enemy breaks through the gaps in the formation, and then the armies collapse into defeat and mayhem. Blood is everywhere; the wounded bodies and the corpses pile over one another and are trampled by both the advancing and the retreating armies. After considering the setting and the conditions of the battles, Hanson (1989: 141) concludes: "Among the Greek armies, the attack resembled more the rush of an armed mob than the march of disciplined soldiers." Plan, order, and discipline are quite outside the regular range of Greek military experiences.

Posssibly the recognition of the importance of having a disciplined plan only materialized after Themistocles' victory in Salamis. One trait distinguished Themistocles' strategy: it required a collective plan, which, in turn, demanded the persuasion of the Athenians. Therefore, Pericles should be seen as a leader who derived his political actions during the war from Themistocles' example. The comparison between the civic and political qualities of both men is inevitably made (Romilly 1963: 192; Ober 1998:

81), with one significant difference: Pericles wants more than a mere military victory over Sparta and its allies.

Pericles' war plan has a diplomatic and persuasive goal: it intends to demonstrate to the Peloponnesians that fighting against Athens is futile. The consequence of coming to this understanding is that the Athenians and the Spartans must build a lasting peace. It is an altogether new and complex way of solving the conflict for the generations to come. Nevertheless, if that was Pericles' objective, the plan was flawed. Kagan (2003: 52-54) points out the misconceptions that diminished Pericles' possibility of success. Even if the Athenians fully understood and totally approved the plan, the Peloponnesians, whose interests and viewpoints about the conflict were completely different, still had to be convinced. Essentially defensive, in spite of the incursions into the enemy's territory and the constant engagement of the Athenian cavalry, Pericles' idea of isolation within the Long Walls does not demonstrate convincingly the might of the city's military resources. At its best, it bothers and irritates the enemy; but the Spartans can easily feel, as it seems to be the case, that they are winning the war against hopelessly coward enemies. At its worst, the result would be an endless and wearing conflict, as it turns out to be. Moreover, Pericles did not take into account the growing resistance among his fellow-citizens to his proposed plan, nor could he foresee the devastation of the plague. In fact, his strategy is a disaster of immense proportions, an impasse that would corner him defensively. As Ober (1985a) maintains, after Pericles, the Greeks would come to reject defensive strategies as synonymous of catastrophe and shame.

Apparently, the Athenians did not comprehend fully the overall design of Pericles' plan. During the first years of the war, until his death, Pericles would have to exercise constantly his capacity of persuasion, or maneuver to avoid the meeting of the assembly. Precisely at the moment when he falls sick, after seeing his two sons dead as consequence of the plague that his strategy of isolation caused, Pericles uses his prerogative as a *strategos* to request an assembly in which he presents arguments supporting his plan. His prestige is not the same as it had been. Under the pressure of some citizens who demanded peace with the Peloponnesians, Athens is about to send an embassy to the enemy. Pericles has to oppose this move. He feels that Sparta would understand the request for peace as a sign of weakness; and this is exactly what happens. The embassy returns to Athens with nothing to offer. Perhaps the Spartans will insist on the same demands that triggered their ultimatum. Or, probably, they will request the freedom of any city that would wish to break up with Athens' leadership. The chain effect of accepting this last request would be disastrous. In practical terms, the Athenian empire would end, at the same time reinforcing the image that the Spartans were responsible for liberating other Greek cities. Spartan power would spread across the Hellenic world. Pericles must have perceived that this was the propitious moment to defend his strategy. When peace fails, the war must go on. The end of the Athenian empire means rendition and defeat.

Pericles complains to the assembly that the frustration and rage against him are unfair and unreasonable. All approved the decision that leads to the present state of things. Collective decisions are the ones with greater virtue and wisdom. It is through a wide consensus that one reaches a kind of fortune, which is greater and more vital than the individual welfare. According to Thucydides (2. 60), Pericles argues "that when the whole state is on the right course it is a better thing for each separate individual than when private interests are satisfied but the state as a whole is going downhill." Gomme (1956b: 167) evaluates that this sentence is the most artificial and least convincing in all of the speeches recorded in *The History of the Peloponnesian War*. Therefore, if Thucydides' addition does not express what Pericles in fact said, it is a judgment on the part of the historian, who in hindsight sees Pericles' decease as the cause of a political void, where demagogues, opportunists, and other public figures thrived without the brilliance and vision of the most important Athenian leader (*protos Athenainón*). One of Thucycides' central contentions in his historical narrative is indeed the attribution of failure to the ones coming after Pericles' death. The outcome of undefined future policies will bring the weakening of the Athenian state and with it the decadence of its empire (Ober 1998: 93).

Pericles' statement about the primordial importance of the state over the individual must be understood as more than a criticism of the Athenian citizens who forgot everyone's share of responsibility in the conduction of the war. Pericles defends himself asserting that he just expressed what was better suited to the interests of the community. He is the same as ever, incorruptible and a visionary. He loves the city and wishes nothing but the best for Athens. Now the audience has changed and is faltering in the face of the misfortunes of the plague. Fear and frailty springing from individual frustrations are not worthy of the collective greatness of Athens. The city has built an empire. Her reputation and prestige depend on each of her citizens.

What is presently at stake is the survival of everyone. It does not matter if other Greeks view Athens' rule as tyranny. There is no other choice (Thucydides 2. 63). The loss of the empire will bring deprivation and destitution to all. It is necessary to preserve the conquests of Athens. Otherwise, one will experience the dissolution of the *polis*, and this is not in anyone's interest. It will be a collective ruin. More than ever, in a moment of crisis and despair, one must think of common interests. The luxury of political apathy is too costly. Chance brought the plague that caused immense suffering, but the remedy for present afflictions is to resign oneself to the misfortunes of fate. Nothing is more Athenian than that. No one has the right to betray the ideal by which one lived up to until now.

At some point in his last speech to the assembly, Pericles boasts about Athens' superiority in the sea. The current devastation of the agricultural districts will be compensated by everything that naval advantage allows. The empire will expand, diminishing the losses. It will be for everyone's benefit. Athens cannot escape her vocation. Here, Pericles does more than simply throw a hyperbole in the midst of impassionate words. Whoever heard what he was saying would certainly guess that his reference to the Athenian naval power is also a praise of its political order. It is too a vindication of what he had done. Deepening the democratic legacy of Solon and Cleisthenes, Pericles has shaped a more inclusive and more collectivistic democracy in Athens.

The outcome of all this is that the Athenian citizens must be fully committed to their city, and that means all of them, not just the upper classes, bu t also the lower-rungs of society, so many times disparaged as the naval populace (*nautikos ochlis*). The wealth and power of the Athenian aristocracy, owner of gardens, parks, and other elegancies, are not what made the city an empire. Social parts ought to behave thinking of the welfare of the whole. It is not enough to act from individualistic interests and then dream that the greatness of the city will come into existence. The part is always smaller than the whole. However, the individual part is, in many circumstances, incapable of visualizing ing the totality. One must have t he intelligence to recognize that all citizens created the Athenian empire. Athenian culture and education forged a great empire. The citizens of today must preserve what took them to where they are now: "It is for you to safeguard that future glory and to do nothing that is dishonorable" (Thucydides 2. 64). Do not send embassies to Sparta; surrender and rendition are not options. The alternative that Athenian culture shows to her citizens in moments of distress is to face calamity with a clear mind and to react immediately to adversities; for a city and an individual, this is real strength (Thucydides 2. 64). Through Pericles' words, and in Thucydides' testimony, culture appears to be ever-present in the combat scene.[7]

Chapter 2
Coordination

For those who witnessed the Spartans – under the command of the general Lysander – and their Persian allies enter a helpless, besieged, and starved Athens in March 404 B.C.E., the spectacle must have been the embodiment of a lingering fear that had haunted the Athenians since the war broke out a little more than 27 years before.

The majority of the assembly had accepted the harsh terms of peace. Athens's Long Walls and the walls of Piraeus – which had protected the city, and allowed Pericles to propose the Athenian strategy of isolation during the early days of the Peloponnesian War – would be destroyed. The Athenian fleet – once the source of the city's imperialistic power – had to be reduced to the number of ships that the Spartans permitted. Athens's empire would be dismantled; all colonies would be free from her control. The oligarchs in exile – fierce opponents of democracy – would return to the *polis*, thus tilting the balance of power. The present political system would be abandoned and the ancestral oligarchic constitution would be restored. And Sparta would ultimately dictate Athens's foreign policy (Kagan 2003: 483). Everything that the city represented was simply wiped away, but it was still a better fate than what most Athenians feared: the total destruction of the city, and the transformation of her citizens into slaves of the Spartans, if not the complete extermination of the people in the hands of the iron-hearted, man-slaying Lysander.

The surrender of Athens was more than the downfall of a city and the demise of an empire. It meant the threat of extinction of the Athenian democratic way of life. However, one year after the surrender, in 403 B.C.E., under the leadership of Thrasybulus, the Athenians rebelled, fought, and defeated the oligarchic regime of the Thirty Tyrants faithful to Sparta. Democracy was then restored, and with it, a social practice that is reserved for speeches, indeed the public exchange of signs, the role of a positive tool in the deliberation of issues that are part of the public sphere.

Whether in the ancient or in the modern world, democracies rely on a non-hierarchical model of social life, where all citizens – but realistically whomever is deemed one of them – are equal. To these free citizens is bestowed the right of delivering free speeches in direct confrontation with opposing viewpoints. Democratic political systems begin with the recognition of structural divisions within the community, and move toward harmonizing the antagonistic forces through the search for a just and fair decision. This is the principle of *eunomia*, the consensual harmony of conflicting interests, that for Solon, the creator of the Athens's political culture, would constitute not only the order of the *polis*, but also the divine "peace and harmony of the whole cosmos" (Jaeger 1943: 141). The Greek reader of Solon's poetry would know that when he referred to *eunomia*, he was talking about one of the *Horae*, the goddesses of the seasons. According to Hesiod, the *Horae* were three goddesses: Dike, who personified justice; Eirene, peace; and Eunomia, the good order. The union of justice, peace, and good order was the foundation of the state. While dressed in mythological robes that would make Solon's vision of the city acceptable and persuasive for the Athenians, the question of the good order was directly linked to the establishment of social conventions (*nomoi*) capable of ensuring that the ideal of justice, peace, and good order was actually lived in Athens. Human beings were responsible for achieving the cosmic ideal embodied by the *Horae*; if not, they would create a world of misery and pain for themselves. That project of social responsibility stirred Solon, and what he did – or at least was attributed to him – was to draft a set of new laws, the *seisachteia* (the "shaking-off of burdens"), of such a scope that they would transform Athens from an archaic society into a commercial state and eventually a democracy. The *seisachteia* cancelled all debts, which liberated a class of citizens from servitude (Aristotle, *Ath. Const.,* VI, 1*)*, and lifted the ban on exporting products other than olive oil. The liberation of the poorest Athenians and the encouragement of the pursuit of wealth through trade with other cities had an enormous impact over Athens. Solon's constitution – indeed a set of social conventions – was fair, conducive to peace, and fashioned good order. All of this is now history, but what is particularly interesting for a theory of social representation is the acknowledgment that the good order is achieved through the implementation of correct social conventions that can well be the upset of older and more unjust constitutions. Without the proper conventions, the social ideal of justice, peace, and the good order are unattainable.

While it is true that the procedures of Athenian democracy were the effects of a later reaction to actual events – namely, the tyranny of Pisistratus – it is also undeniable that if Solon had not put forward his vision of the ideal Athenian state, perhaps the city would have difficulty in committing herself to a democratic political regime. Divisions and factions are constitutive of any democracy, but what makes the democratic state politically feasible is the contractual acceptance that one's point of view may not be the chosen public policy. After the votes are counted, the democratic citizen must abide by an implicit social contract: the decision of the majority is *eunomia*. *Eunomia* is the reaching of a consensus after dissent. Dissent disappears through persuasion. Persuasion and the contractual submission to a collective decision are the fullest expression of the good order.

2.1. Democracy, warfare, and the political system

Not all Athenians accepted the democratic way of reaching public decisions. Although incomparable to what is nowadays named parties, in Athens, two factions were involved in strife about whether democracy was the best political system to make decisions that would affect the whole city. The city was split into blocs: the oligarchs and the democrats. The democrats upheld the presumption of wisdom in decisions made by all citizens present in the assembly, whereas the oligarchs felt that democracy should be abolished for it betrayed the most solid aristocratic Athenian traditions. The oligarchs argued that tradition and respect for social hierarchy provided the good order. Democrats responded that conventions could be changed if the change was for the best; and democracy had been proven the best political system. The discussion could have gone on endlessly, but, like so many customs in the ancient Greek world, up until the fall of Athens, democracy had survived because of the continual success of democratic decisions in warfare.

It is correct to assume that the Athenian victories during the Persian wars were not the sole and direct effect of democracy. Democratic decisions could boost the passion and the commitment of an army, but surely, that was not enough. Other factors may have played a decisive role in the military triumphs of Athens. The Persian miscalculations could have brought the defeat of their numerically superior force. It could well be that the Greek hoplites were better individual soldiers than the Persian infantry or that the hoplite strategy was bewildering for the Athenian enemies, and that Athenian triremes were sturdier and faster than the Persian ships during the battle of Salamis. Although such considerations sketch a far more complex and perhaps historically truthful understanding of the Athenian military achievements, it cannot be dismissed that for an ancient Greek the stunning successes of his city had in fact occurred after democratic resolutions and that had to be the confirmation of the superiority of the democratic political system. In Homer's *Iliad*, the Greek army was depicted as non-hierarchical and some their great heroes – like Nestor and Odysseus – were *men of winning words as sweet as honey*. Moreover, examples could be extracted from reality. Annually, the assembly chose, though direct vote, the citizens who would be the *strategoi.* If the choice was not correct, if the chosen for this important position made mistakes or disobeyed the city's mandate, the assembly would convene to punish him – severely with exile, death, and the confiscation of property – and then choose another general. The procedure had been suited to the needs of the city.

Before the battle of Marathon, the Greek commanders had discussed openly when to face the Persians. Herodotus, in Book 6 of *The Histories*, tells of the division half-and-half amongst the Athenian generals just before the battle. Some felt that they should not face the Persians: their force was too small to defeat such a numerous enemy; others were of the opinion that they should take the chance and fight. At least they would have the possibility of wining. A *polemarchos* (a nominated war commander) named Callimachus cast the deciding vote after listening to an impassioned speech of Militiades, who urged them to fight. The Persians lost 6400 soldiers; the Athenians, 192. The battle of Marathon turned out to be an improbable success: an army of 10,000 Greeks crushed a mass of Persians probably double its size.

Herodotus recorded in Book 8 of *The Histories* how Themistocles had transformed the frightening omens of the oracle of Delphi into persuasive argumentation, managing to evict the citizens of Athens from their *polis*, avoiding infantry fighting with Persians, and finally confronting them with ships in the strait of Salamis. When asked by the Athenians about what the future would held for them now that the Persian invasion seemed imminent, Aristonice, the Priestess of Delphi, pronounced a terrifying prophecy, announcing that the Athenians should fly to the world's end, and that "the headlong God of War speeding in a Syrian chariot shall bring you low." After reading the transcript of the oracle, despair spread among the Athenians, who decided to send new envoys to Delphi, with olive branches for Apollo to supplicate for an outcome not so disastrous. The oracle's prophecy changed little, but the change could be relevant; it said, "[T]he wooden wall only shall not fall, but help you and your children." The Athenians were divided about what the line meant. Some saw it as a reference to the Acropolis, which was fenced in the past with what could be a wooden wall. These people understood that the city should face the Persians from there. Others thought that the wooden wall was a fleet of ships aligned for battle. The prophecy ended with enigmatic lines: "Divine Salamis, you will bring death to women's sons when the corn is scattered or the harvest gathered in." Some of the Athenians who read the wooden wall as ships saw these lines as foreboding defeat in Salamis. The strategy of fighting the Persians at sea had few chances of being approved in the assembly. Then Themistocles twisted the sentence around and claimed that if it were an omen of defeat for the Athenians the expression would not be "divine Salamis." Defeated Greeks would naturally say "hateful Salamis." If that was not the case, the oracle was really predicting the defeat of the Persians in Salamis. Persuaded, the Athenians prepared for the battle in which they would be victorious. Once again, a democratic decision ensured military success; but now, with the end of the Peloponnesian War, and the defeat and submission of Athens to the Spartans, the picture had changed.

When, in the dialogue *Phaedrus* (267 a–b), Socrates described the teaching recipes of the sophists, and chastised Tisias and Gorgias of Leontini for having taught that "probability deserves more respect than truth," it is hard to think that the remark was meant just for these sophists. The resonance of the criticism was certainly wider. Putting this criticism in the mouth of Socrates, Plato had in mind other sophists, and he was addressing the Athenian citizens who had been recurrently seduced by probable arguments flaunted at them. Such was the case of the popular approval for Pericles' ill-fated plan of isolating Athens within her walls to combat the Spartans, and more recently the support for the disastrous Sicilian expedition, approved by the assembly in the winter of 416–415 B.C.E., after two Sicilian cities, Segesta and Leontini, sent ambassadors to ask the Athenians to help them against the conquering plans of the Syracusans. The argument of the Segestans and Leontines was based on logical probability. If unrestrained, the Syracusans would conquer the whole of Sicily and they would thicken the Peloponnesian forces, which could possibly be deployed to crush Athens.

During the assembly to evaluate the Segestan proposition, the debate was heated. The respected, pious, wary, and older general Nicias opposed the idea. According to Thucydides' *The History of the Peloponnesian War*, the cautious *strategos* said that the Athenians should not be involved in adventures so far away from their city. The consequences were unpredictable. The city would create new enemies. He also alerted, without directly naming the accused, although everyone present knew that he was talking of Alcibiades, that someone in the assembly would certainly approved the proposition of sending ships to Sicily, not because of his concerns for the welfare of the city, but because he had eyes on personal glory and wealth. A protegé of Pericles and a member of Socrates' inner circle, the dashing Alcibiades responded that to help the Segestan was their duty as an empire. "This is the way we have won an empire, and this is the way all empires have been won" (Th. *Hist*. 6. 18), by helping vigorously all of their allies.

To sit inactive is not what an empire should do. Empires are active forces: they should not only defend themselves, as Nicias had suggested, but also take measures to prevent future attacks. The speech was not grounded on any assessment of the conditions and consequences of the expedition. It was peppered with generalities. It was tailored to impress the audience, and looked as though it were made directly from the sophistic recipes widely taught in Athens. Moreover, it was similar to the one that Pericles had delivered before, when proposing his plans to fight the Spartans: the future of the Athenian Empire was at stake.

The Athenian assembly had approved both propositions – Alcibiades' and Pericles' – because the plans appeared reasonable, though potentially inconsequential, at best merely probable. The social tragedy of accepting Alcibiades' proposition was even greater. The Athenians were sending a military expedition to a place that they did not know very well. According to Thucydides' *History* (6.1), the assembly was ignorant of the size of the island, and the number of its inhabitants. It was like getting involved in another war of the same the magnitude as the one still raging against Sparta and its allies. It was sheer recklessness. The choice of the *strategoi* to conduct the campaign was an example of such thoughtlessness. The three chosen generals were Nicias, who had been against the proposition but could not reject the appointment for fear of the charge of being unpatriotic; Alcebiades, whom the assembly never fully trusted and who was a political enemy of Nisias; and Lamachus, a general reputed to be a bully who eventually died heroically fighting the Syracusans.

The lack of consistency and uniformity in the joint appointment of the three *strategoi* was in itself alarming, especially if one considers that as the preparations for the expedition were taking place, a rumour began in Athens insinuating that Alcibiades was a participant of the sacrilegious plot that destroyed the statues of Hermes. Although under such a serious accusation, Alcibiades was never demoted from the role of *strategos*. He eventually eloped to Sparta to help the enemies of Athens. On top of all that, to dissuade the Athenians from what he deemed to be a risky enterprise, Nicias suggested that they should send to Syracuse a much more powerful force than the original proposition of 60 ships. His hope was to cool off the assembly's enthusiasm, and yet the Athenians thought that it was great advice. Now, the city was not only courting disaster, but was plunging as well into a folly of even greater proportions. In the end, when the Sicilian expedition was over, the Athenian casualties were huge: 3000 hoplites, 9000 tétes – men of the lower classes who usually furnished the oarsmen of the fleet; and thousands of metics – a special class of individuals composed of foreign-born residents of the city (Kagan 2003: 327).

Democracy had by then lost its aura of wisdom. The opponents of the democratic method of making public choices were growing. Democracy was seen as mob-rule. If led in one direction or another, the assembly could turn against anyone, and even forget its own previous decisions. The assembly was an unpredictable, fearsome, and unfettered force. Even when precisely informed of the casualties and the result of the Sicilian expedition, the Athenians refused to believe in what they were hearing. The previous possibility of victory that appeared to be so certain was not to be falsified. The Athenians pounced on the public speakers who had been in favor of the incursion into Sicily. The Athenian mob was enraged at their soothsayers and oracles. The citizens forgot that they had voted for the expedition (Thucydides 8.1). Although with a depleted army, a fleet in shambles, and empty coffers, the Athenians decided not to give in. They would fix their problems. The insane war with the Spartans kept on going for almost a decade.

It was inevitable to wonder what had gone wrong. Was not consensus the criterion of the good order? The expedition had been decided consensually: what had been the problem then? Thinking about this perplexity, Plato pointed out that the problem was that the city could not wish for the good order if at the mercy of untrammelled ambitions and political passions. Strong, passionate, contradictory, and irrational speeches could easily infect a crowd. The whole democratic process – the city itself – was doomed when social resolutions were based on persuasion only. Much more than that would be necessary. Under the influence of passions, the crowd would be incapable of judging the long-term consequences what was presented to them. Unscrupulous demagogues and unprincipled public speakers were a deadly combination. People like that poisoned the political body of Athens (*Republic* 8, 564 b-c): for their sole ambition was to accumulate and maintain power. Nothing will stop them: they do not deserve to live in the community. When Plato declared that he could not tolerate the idea that all opinions were equally valid: Truth exists because, contrary to what Protagoras proclaimed, lies can be told (Euthydemus 286c) Although not for the old sophist, now that Protagoras was dead, but for all the impious and atheists, in fact anyone who mocked the possibility of a divine centre irradiating truth, for diviners, fanatics, impostors of all kinds, dictators, demagogues, generals, whomever practiced the tricky arts of the sophists, for all of them one of Plato's characters in *The Laws* – the Athenian – said that they deserved to be in a special prison. This prison was quite different from two other jails: one where the common persons would be locked, and another that would function as the nocturnal house of correction. The detention of such nefarious anti-social characters would be in an isolated incarceration at the heart of the country, distant from all, in a most solitary place (*Laws* 10, 908 a-b). The city had to protect itself against the possibility of being undermined from inside. Sadly and hopelessly, democracy undermined itself.

The fatal error of Athenian democracy was to live by the conjecture that consensus is an absolute value. Consensus could not be an indication of justice, or the incarnation of the good order, if justice and good were completely undefined. Without a firm foundation, consensus really does not matter: more important than common agreement would be the submission and the coordination of personal and selfish interests to a system of values that should arise from a precise body of knowledge composed of answers to questions about what justice is, what peace is, what the good order is. Rhetoricians – like Protagoras and Gorgias – who denied that there were no answers to such questions had no place in Plato's *polis*.

2.2. The contrast of nature and conventions

What is known of Protagoras' life and works is scant. Other than an early and amusing Platonic dialogue narrating his encounter with Socrates and their sparring during one of Protagoras' visits to Athens, what survives about the elder sophist is a collection of ancient fragments (see Sprague 1972: 4–28) covering apocrypha and doxography. With variations, the collection of fragments repeats his most famous statement: "of all things the measure is Man." The celebrated human measure – the *anthropos metron* – is less a prideful statement than an inference reached from the discovery of human limits. Humans can only know what is human: their knowledge of the gods is non-existent. Humans cannot say they know the gods, much less that the gods can be known. Divinity cannot be ascertained. Because seeing and knowing are intertwined in humans, the knowledge of gods is impossible. The divinities are invisible, and not only that: while the gods are eternal, humans are finite. Human life is too short to glimpse any remote trace of infinity.

The recurrent accusation of Protagoras' atheism is not adequate. He did not deny the gods. He could not do it. He simply stated that humans could not know them. For Protagoras, it was impossible to know if gods are infinite or immortal. That is an uncertain hypothesis. While not strict atheism,

Protagoras' subversion is nonetheless damaging: men are the measure of the divine. Enraged by the idea that the divinities do not measure men, but rather that men measure the divine, Plato counter-attacked: "now it is God who is for you and me of a truth, the 'measure of all things' much more truly than, they say, 'man'" (*Laws* 4, 76 c–d).

Devoid of a divine foundation, incapable of finding a true unifying point for them, human beings are thrown into a tumultuous whirlwind of conflicting interests and viewpoints; they dwell in contradictions. Each human being has a distinct and very selfish tendency to behave in society. That is a natural law, and yet, is that correct? Is it possible to live in a world like that? If, for nature, it is inevitable the search for pleasure and satisfaction, then each individual looks for what seems good and that search is unbridled. It is directed toward not the purpose of a common good, but to the immediate accomplishment of personal welfare and selfish satisfaction.

Because of humanity's confinement to a contradictory nature (*physis*), even the human evaluation of the divine is unstable. It is inadequate to conclude that either the gods are or they are not. Nature expresses itself through contradictory dualities, and it is not viable to decide about such multiplying contradictions. Barred from reference to a divine normative standard, human beings can only guess if something is a possibility. Humans cannot have complete certainties.

It is useless for human beings to look at nature for attainment of certitude. Nature is also deprived of an absolute normative standard. The natural norm is circumstantial and reduced to the quest for personal advantage, and that means nothing more and nothing less than the selfish search for the useful and profitable to each specific individual viewpoint at a certain moment. In nature, the viewpoints and the state of affairs multiply endlessly. The number of viewpoints corresponds theoretically to the number of individuals. How can one make a definitively true decision in a godless world?

Humans must act, but they do it blindly: no one is able to know for sure what is right or wrong. Protagoras then affirmed that relativity and fluidity are the traits of nature. Contradictory attributes do not exclude one another: they co-habit in each natural being. The water of oceans is, at the same moment, a pure and an impure environment, a wholesome and an unwholesome habitat. Fishes will live in it, but men die if drowned. Hot is developed into cold. Birth turns into death. Is there, in a world such as this, the faintest possibility of having a uniform perspective, or are humans caught vulnerably in just a complete entanglement of knots? Is there an absolute truth or is there nothing but contradictory viewpoints? Does truth exist at all?

Although Protagoras appears to uphold a staunch subjectivist stand as far as knowledge and morality are concerned, that is not quite his position. So if everything has an opposite, what is the opposite of nature? Is it subjectivity? The opposite of nature could not be anything with extreme singularity. The antithesis of natural singularity is thus social generality. Nature abhors conventions, while humans live by a set of arbitrary and conventional rules. Nature is infused with passion, desire, intensity; human laws and conventions are the opposite of nature: they are not singular, but offer collective norms to humans so that they can regulate and subjugate natural passions. *Physis* and *nomos* are in stark contrast.

Protagoras was a thinker whom Plato assailed constantly, because, for him, the old sophist's teaching was addressing incorrectly a crucial issue for Athenian society. His theories aggravated even more the already unstable social life of Athens. In an assembly deciding the future the city, how should the individual citizen behave? What should he consider more important: his self-interest or the troubles of the community? How can one be sure to recognize the voice of the community? Was not the approved decision just the personal viewpoint of someone who had better rhetorical skills, who conveyed his ideas more forcefully? It is as though the criterion of force that decides conflicts in nature was never superseded. Either the law of the *polis* is the supreme norm for human life, in perfect match with divine order of the existence, in such a way that humanity and the citizen are one and the same thing, or else the laws of the *polis* are in direct contradiction with the divine and natural norms. In that case, it is perfectly legitimate to disregard them and act according to one's most selfish interests.

Surprisingly, from his radical tenets Protagoras did not come to a radical conclusion at all. While humans are incapable of discovering an absolute and timeless norm, they can perceive the notions that are consensually held in their communities. That was far more important than the individual's selfish viewpoint. Protagoras professed then that of all possible speeches the stronger ones are based on traditions and common cultural values. Not only such speeches are more attractive to audiences, but they also carry greater force. It is correct to take up the ruling ideas of the community and parade them in speeches. A citizen should accept the revered gods in the community, not because he believed in their existence, or because of his fervent creed, but because traditions dictate so. His only possible attitude is to accept the cultural costumes of the group. Truth is turned into convenience and conformity.

If cultural habits are the foundation of societies, social life rests on quick sands. As time unfolds, costumes change, and looking around, it is easy to distinguish the differences of cultural customs between the *polei*. So why should the Athenians uphold their traditions and cultural habits? Why not adopt the Spartan traditions? It could not be just because it has been like that always. In fact, it was not always like that: Solon's draft of a new constitutional order showed that conventions are arbitrary and therefore can be altered. Under the influence of the sophists like Protagoras and Gorgias, the new generation of Athenians was taking a cynical and ruthless approach to their society. The health of the *polis* was being sapped away.

2.3. To have a civic morality

The most illustrious and revered of the ancient sophists – Gorgias of Leontini – knew very well that speeches could be dangerous instruments. The correlation of seductive magic and rhetoric was well established in the Greek mind (Romily 1975). In his extant speech *In Praise of Helen* (see Matsen, Rollins, and Sousa 1990: 34–36), Gorgias compared speeches to drugs; they transmit divine sweetness, induce pleasure, and reduce pain. Like drugs, speeches have opposing effects; they can end disease, but also they can end life. These remarks are part of his defence of Helen, who was a puzzle for the Greeks. Did she run away with Paris? Was she abducted? Should one pity her? Should she be abhorred? From "whatever she did, persuaded by speech, seized by violence or forced by divine necessity, she is completely acquitted" (*Pr. Helen*, 20). Persuasion, the insanity of love, and to be carried away by a powerful divine force deadens the personal will: they are major overwhelming forces.

If Gorgias of Leontini knew that, why did Plato make him declare in the dialogue *Gorgias* that rhetoric is a neutral skill? Gorgias proclaimed that rhetorical skill was similar to learning how to fight or to use a weapon in a combat. Suppose that someone has learned how to box and becomes a fearful boxer, and then goes on striking his mother and father, is it the instructor's fault? Is the art of boxing to be blamed? Neither the teacher nor the craft should be censured, but "those who make improper use of it. And the same argument applies to rhetoric" (*Gorgias*, 457 a–b). The thought was disingenuous. There, in the Athenian market place, after coming from a party, Gorgias, two of his disciples, Polus and Callicles, met Socrates and Chaerophon. The encounter grew into a verbal combat between Socrates and the sophists, Gorgias, Polus, and Callicles. Plato's dialogue was divided into three sections, each reserved for the confrontation of Socrates with the three main characters. The section that set Socrates against Gorgias was a discussion about what rhetoric is. The debate with Polus moved around a moral paradox: is it better to commit or to suffer evil? With Callicles, Socrates discussed the definition of a good life. In each of the sections, the moral deficiencies of rhetoric and sophistic education were brutally exposed. The only dramatic role of Chaerophon was to serve as a go-between the contenders, for he was also a friend of Socrates. The friendship of Chaerophon and Socrates was solid: in his trial, Socrates called him a friend and publicized that, during one of his visits to Delphi, the oracle had told Chaerophon that he was the wisest man in Greece.

Gorgias may have acknowledged the dangers of rhetorical teaching, and he may have tried to find an excuse to exonerate himself from being responsible for what he taught. However, in ancient Greece, an educator was judged primarily by what his pupils did. Such a responsibility was not easy to dismiss. What pupils did was the clearest effect of the educator's work. In Socrates' trial, it was obvious that accusation of corrupting the youth seethed the Athenian resentment against at least two of Socrates' disciples: Critias, who had participated cruelly and hatefully in the governance of the Thirty Tyrants, and Alcibiades, who betrayed the city to the Spartans, and was seen as conspirator against democracy. During the meeting of the three sophists with Socrates, Polus and Callicles personified the wicked effects of the kind of training that Gorgias conferred upon his students. It was made clear how ruinous that education was for the city.

Like a brisk, exuberant colt with no reigns, Polus more than once budged in the discussion of Socrates and Gorgias. At some point, he ended up cynically declaring that the happiest man in the city was the tyrant, for he could do what he pleased, and act according to his unencumbered will. He is free to kill, to exile, to follow his own pleasures (*Gorgias*, 469 c–d). Polus confessed that he would act that way if he had the power. Socrates then wondered: if he had that chance, would Polus be a happy man? Could anyone be happy doing wrong things? Was it hard to see that this kind of life is infused with shame and wretchedness?

Gorgias apparently never taught a moral lesson to his disciples, not to Polus, not to Callicles. This young aristocrat, a mysterious character, who may have been a fictional creation of Plato or a distillation of the traits of Athens's younger generation, countered Socrates' paradox. Callicles proposed that what was called justice was the manifestation of the force of superior individuals. Such individuals were the natural rulers, despite the resistance of human laws (*nomoi*) created to curb the will power of these people. Social conventions were made to protect the mediocre. They were created to set limits to the ones who will naturally be the strongest. Tyranny was the most natural and desirable political state of affairs. For him, despotic pillage deserved admiration.

Amazingly enough, both Polus and Callicles looked up to despots as their political models. If the goal of the educator was to perfect the disciple, then sophistic and rhetorical training only brought corruption. While teachers left instruction with their pockets loaded with the student's money, the pupils would never find happiness. How could they? Happiness only occurs if the individual is perfected. It is an indication of unhappiness to have our nature debased. And that surely happens when the opportunity of perfection is lost. Rhetors and sophists threw away the occasion to make constructive changes in Athens through the education of her youth.

Looking at Plato's dialogue, and considering each character in *Gorgias*, it is both interesting and relevant to examine the sequence of characters that dispute with Socrates. The order of characters sets up a sense of progression. Moving from Gorgias of Leontini to Polus and finishing with Callicles, Plato exhibits the transition from the teacher – indeed the cause of it all – to the students who were visible effects of what was taught to them. Socrates' presence and consistency offered the alternative to Gorgias' narrow intellectual latitude and moral capacity. During their interaction, Socrates' steadiness indicted Gorgias' willingness to probe into the foundations of what he was teaching. Gorgias was unfit to concentrate on the fundamental issues that would have an impact over the city's political present and future. If Gorgias was incapable of answering the Socratic inquiry on the nature of his craft ("what is rhetoric?"), then how could he face thornier problems? That definitely disqualified Gorgias, the teacher. And how about other questions; how would the great sophist present answers to "What is justice?" Moreover, how could, from his instruction surface a reply to "What is the good order?" By his own admission, and quite similar to Protagoras, Gorgias advocated the hopelessness of knowing. Sextus Empiricus (*Against the Schoolmasters* VII 65, in Spraguer 1972: 42) described the progression of Gorgias' ideas in the treatise *On the Nonexistent* or *On Nature*, as composed of several headings: first and foremost, nothing exists; second, that even if something exists it is beyond our comprehension, and third, that even if something is apprehensible, still it cannot be communicated to other people. Nothing was taught, for nothing could be known. The teaching was reduced to the instruction of empty forms of presentation, coupled with the voided means of arguing through questions and answers. Polus and Callicles were offshoots of such training.

Nonetheless, at close range, Polus and Callicles were at odds with each other. Polus understood the quest for power as a result of a trite generalization: everyone craves power. Callicles, however, conceived power not as a pathetic personal wish: it was part of the natural drive; the roots of Callicles' conception of power went much deeper: its roots were in necessity. Power struggles were at the core of nature, which meant that they would never be eliminated. Yet, despite their immediate diversity, Polus and Callicles were related to one another in such a way that they seemed to be gradations of the same spectrum. If Polus was a real person and Callicles a fictional character, what Plato was doing was pointing out that Athens was on the verge of being guided by ruthless and insensitive people. Polus was the close present and Callices, the gloomy future.

2.4. Starting with signs

Protagoras and Gorgias taught that nothing exists beyond the signs exchanged in the assembly. If they were right, there would be no certainty. All political messages were nothing but dazzling verbal displays, merely fickle language. Gorgias preached that an impressive discourse was just about all that one could expect. The more ornamented, the flashier the display, the more persuasive would sound the delivery of the speech sounded. Nothing could be known. Nothing could be truly communicated. The cynicism of Polus and Callicles matched Gorgias' scepticism perfectly, but was Gorgias as sceptical as Sextus Empiricus portrayed? The argument of Gorgias was so extreme that it may well be that the reflections of *On the Non-existent* were just a humorous parody of Protagoras. No one can be sure of that, but the cynical and sceptical views of political power were circulating in Athens. The young Athenians took them seriously, for the effects showed in Polus and Calicles.

On his part, Plato obviously took the issue earnestly. His reaction was less motivated by the notion of producing an intellectual alternative to the abstract gambolling of Protagoras and Gorgias than by the urge of providing the cure for a disease that in his view was infecting the city. Although the subjects of truth, representation, and moral ground can be seen speckled throughout Plato's dialogues, it is in *Cratylus* – one of his works most notorious for posing difficulties of interpretation[1] – that he will refer to the issue of the foundation of signs from the strict standpoint of language.

The characters of the *Cratylus* – Socrates, Hermogenes, and Cratylus – are discussing the foundations of naming. How is it that a sign is applied and referred to an object? While this is what the characters in the dialogue appear to be talking about, it would be hasty to state that the theme of the *Cratylus* is exactly language. Robinson (1955: 222) observed that the term used in the dialogue is other than the proper name. *Onomaton* in Greek is quite close to the notion of name, although it may be more exact to use the term *word*, as long as one has in mind that it can refer to adjectives, but rarely to verbs, and even more rarely to prepositions and conjunctions. Thus to think that the subject of the *Cratylus* is strictly language would be a distortion. In Plato's case, an abstract theme in a dialogue is always related to a pressing political and social issue. Language could never be the sole object of Plato's concerns; language is analyzed because its use can tear the city apart.

Perhaps what Plato was above all doing was investigating if there were an alternative to conventionalist doctrines, so obviously frail to ground representations, and so dangerously akin to Protagoras' and Gorgias' relativistic contentions. The development of the dialogue is clear: it represents a critique to what appears to be the only two alternatives concerning signs and political speeches in Greece and ends up proposing a counter-hypothesis to them. The goal is to transcend the stubborn rivalry of nature and culture. Adopting one or the other doctrine was always unsatisfactory.

With his contention, Plato wanted to bury naturalist and conventionalist theories. If so, what are the characters of the dialogue really thinking about as they interact with each other? Hermogenes invited Socrates to talk with Cratylus. Hermogenes and Cratylus are taking opposite sides in the discussion. Hermogenes maintains that names are applied to objects through conventional processes, while Cratylus, who remains silent for almost a third of the dialogue, emphasizes that names are granted naturally. Cratylus' idea is that for each object there is a correct name. For being natural, the correct name transcends cultural barriers. Correct naming is a universal connection: "there is a truth and correctness for names, which is the same for Hellenes and for barbarians" (*Cratylus*, 383 b). For both Hermogenes and Cratylus, the direct connection of the name and the referent grants truth to the act of naming. They would differ about the cause of the connection: Hermogenes attributes it to social conventions and Cratulus to a natural link. The flickering feeling in reading the dialogue is that Cratylus' naturalist hypothesis is not really the object of Socrates' critique, although he does not agree with either. The principal aim of Socrates is what Hermogenes openly upholds.

For Hermogenes, all names are correct, not just some but all. That is not a product of nature. Its cause is habit. In fact, it is not very clear what Hermogenes means. It could be agreement, convention, stipulation, or customs (Ackrill 2000: 130), not merely in isolation, but combined. Agreement presumes a communal process; convention is a shared set of rules; stipulation is an imposition; and custom really is the legacy accepted as a tradition.

Beyond such quibbling for semantic nuances, in Hermogenes' conception what makes a correct and suitable name is that it depends on its common use in a social group, not universal correctness. Apparently, the opposition of Hermogenes and Cratylus is patent: it is one thing to state that the correctness and suitability of the name comes from the natural relationship of the name and the referent and quite another to state that this correctness comes from whatever – agreement, convention, stipulation, costume – makes *all* in a group use the same name. The dialogue *Cratylus* suggests that conventional and natural naming are mutually opposed. Socrates undoes the impasse, dislocating the argument to another level.

The demonstrative effort of Socrates is to prove that both Cratylus and Hermogenes are wrong. Criticizing them, he reframes the meeting as a discussion about the underpinning of correction and appropriateness of naming (*orthotes onomata*). From this viewpoint, Hermogenes and Cratylus arrive at mistaken conclusions because they did not address the complexity of the issue. Choosing between conventions and nature would result in error: conventions and nature are interlocked in a most bewildering manner. Naming is the creation of rules that makes signs cling to the referred object. Are all rules arbitrary, like Hermogenes appears to affirm? Not all; there are signs that bounce from relationships of cause and effect and are true and appropriate all across the natural world, such as the coils of smoke going up to the sky if a house is on fire. In the same vein, likeness cannot be dismissed as a means of starting up a rule and that again would cut across cultures. It is no wonder that Hermogenes and Craylus never addressed any other type of signal, which was not conventional. Their reasoning went thus in circles. Nonetheless, more important than the type of sign analysed is the consideration of how the advent of rules – whether based on cause and effect, likeness, or conventions – creates the motivation to use the sign. Naming sustains itself by tagging along with the rule. Once the conventions are established through the incessant trading of signs among users, naming looks natural, immutable, and non-motivated. Users name objects after the complex establishment of a correct and suitable name in which nature and conventions blend. Neither Cratylus nor Hermogenes could be right.

When the dialogue takes a stray course, Socrates directs the discussion back to its original topic. It happens initially when Socrates criticizes the implicit relativism of Hermogenes' conventionalist stand.[2] Hermogenes states unambiguously that naming is conventional and arbitrary; it is equally possible to accept that an individual calls a horse "man," while others call men "horses." Both acts of naming are right. Socrates then questions Hermogenes directly: *You would acknowledge that there is in words a true and a false?* Certainly. *And there are true and false propositions?* To be sure. *And a true proposition says what it is and a false proposition says what is not?* Yes, what other answer is possible? *Then in a true proposition there is a true and a false?* Certainly. *But is a proposition true as a whole only, and are the parts untrue?* The parts are true as well as the whole. *Would you say the large parts and not the smaller ones, or every part?* I would say that every part is true (*Cratylus* 385b–d). Now, the inference was obvious, if not only whole propositions are true, but also their parts – their names, their words – then there are true and false names and words. Naming could not be arbitrary. On the two cases that Hermogenes imagines in the first, one would be calling horses under the wrong name, while in the second one would be using the correct name. The contradiction is blatant. Agreeing with Socrates' line of questioning, Hermogenes undoes the characterization of names and words as arbitrary, and thus refutes himself.[3]

Hermogenes' initial proposition is ripped to shreds. He sustains the untenable. It is out of the question to agree with Socrates' thesis that there are correct and suitable names and words and at the same time support the extreme relativism of his initial position. How could one endorse the notion that one can name things following radical whims if names and words exist for the purpose of communication? If anyone could communicate anything with any word at all, communication would be impossible.

For Plato, the dialogue – to the fullest extent of the term – is the very foundation of truth, knowledge, and society. Without the interaction of minds, all sociability crumbles apart. And in society we do talk constantly, although mainly in a non-philosophical manner, trading trite remarks, merely chatting. If Hermogenes' radical relativism were true, dialogues would never occur; and they were happening between Socrates, Hermogenes, and Cratylus. Plato's works are constantly marked by the attempt to demonstrate, using the specific literary art of the dialogue, that truth is never the property of one mind in particular (Weingartner 1970: 9). The road to truth always demands the communion with other minds.[4]

2.5. Exchanging signs

After comparing names and words to instruments performing specific tasks, Socrates asks Hermogenes what the purpose of naming is. Hermogenes does not quite know what to answer, and Socrates insists: *Do we not give information to one another, and distinguish things according to their natures?* Certainly we do. *Then a name is an instrument of teaching and of distinguishing natures, as the shuttle is of distinguishing the threads of the web*. Yes. (*Cratylus*, 388 b–c). Hermogenes, once again, falsifies his own contentions. The purpose of sharing information is incompatible with Hermogenes' naïve relativism. If names and words were arbitrary, the same information could not be shared. Names and words must have a certain degree of precision; otherwise, information would be inaccurate and variable to the point of having no meaning.

Then, Plato drafts a critique to a more sophisticated brand of conventionalism. If instruments are analogous to names, and if both are made to do something, this happens because they did it according to their own nature. The correction and suitability of names and words therefore results from their correspondence to a common form connecting thing and sign, just like the case of a smith that makes hammers following the same form in his mind, though using varied materials, and still the instrument may be good of whatever iron is made (*Cratylus*, 390 a). The same applies when languages are compared, whether in Hellas or in a foreign country – there is no difference, Cratylus tells. For that reason, a name and a word could refer to several individual members of the same class of objects. That is why a name is correct and suitable for a class of horses, and it is true and correct for all its users.

To understand naming only from the perspective of conventions is not acceptable. It entails an incomplete way of exchanging signs. Both participants of the dialogue may even know what is said and may even recognize the meaning of a statement, for conventions can be stable and common to the speakers, but something is missing in this type of communication. One cannot determine if the statement is true or false, and this defaces the whole purpose of exchanging signs, which is to convey true or false information. The only means to satisfy the criterion of truth – a condition for uttering any statement – is to abandon any idea of arbitrariness of the sign, and support the notion that there is a true, correct, and appropriate use of a name or a word.

When Socrates refutes conventionalism, does he accept naturalism in naming? If conventionalism is discarded, does it mean that – right away – Cratylus' thesis is correct? Apparently, Socrates took Cratylus' side when he states the existence of "the name, which each thing by nature has" (*Cratylus*, 390 e). Socrates berates Cratylus' support for the theory that nature produces correct and unquestionable names. He asks Cratylus: if all names are correctly imposed upon the objects, does it mean the name meets that standard of absolute fidelity and therefore the ideal name would be undistinguishable from its referent? Cratylus agrees. If that were true, the naturalist theory would come to an impasse, simply because there is no name that would satisfy fully the stipulated condition of native and absolute likeness to its object. If that were so, the letters of a name would have to be exactly like their objects (*Cratylus*, 434 a); it would never be the case: there is always a difference between the picture and what it represents. Consider painting, for example: pictorial depiction always begins with the pigments covering the painted surface and never from the depicted object, which is left behind in the process of representation. There is no native and absolute likeness to an object. Likeness is approximation and frequently a loss. Cratylus then admits his mistake, and agrees with Socrates' criticism.

2.6. From signs to values

Nonetheless, even if it were possible to have a perfect representation, an absolute coming together of signs and things, one problem still hinders Cratylus' theory. The conception of correct and suitable names only made sense if one conceded the existence of error and falsity. Names and words could not be always true. What assurance do we have that the Demiurge – or the original nominators – gave the correct and appropriate name to things? How can we possible know if, after the original naming, human beings kept on repeating the mistake to a point that we presumed that those were the right names when in fact they were not? Cratylus reacts to Socrates insisting that the original legislators ought to know what they were doing; otherwise, "the names would not be names at all" (*Cratylus*, 436 a). If everything one knows is through the use of words and names, how can one be sure that the original legislators had the necessary knowledge for the correct and suitable naming? Knowledge at all times precedes naming: signs are not self-enclosed territories, without reference to a cognitive norm. Now, Socrates responds to Craytlus: is it not true that everything in nature is becoming and is lodged in a state of everlasting flux, and that this feature of relentless change is everywhere; thus it must have been also bestowed to naming? Cratylus agrees with Socrates. Thus, although the naming may have been correct and suitably applied at the origin, and as names are under the effect of the process of becoming, what guarantee do we have that names are always correct and suitable? If they were once, they could not be anymore. If they were not correct at the beginning, why they should be right now?

The recognition of a potential falsity, without which there is no correction, is quite akin to the necessary admission that there are good, bad, and unfair men in the city, and that there are speeches that are wicked and others that are enlightening, and that there is good and bad rhetoric. The logic is simple: without evil, there is no good; without falsity, no truth; without error, no correction. The scepticism and the extreme relativism of Gorgias and Protagoras are not just bogus. Their refusal to acknowledge the necessity of referring to a set of opposing standards requires more than a logical misstep. Their indifference to values seeds the moral and civic destruction of the *polis*. Polus and Callicles were no more than reminders of what would come about. Plato suggested forcefully: the focus on overwhelming self-interest is a vicious trait that should not exist in the ideal society. Self-interest should be subordinated to the values that coordinate civic and political life. Concerning Cratylus' hypothesis about the universal and steadfast correction of natural naming, Socrates finalized the argument, pointing out that if everything were true, then the discussion about the correction and suitability of naming would be at once punctured. Why argue about that? The condition of truth and correction is the admission of a potential falsity. No one is correct if following the all-embracing and conclusive dictates of self-interest; that is not right all the time; and for the same reason, not all names are correct, suitable, and true.

Although truth is established as the fundamental condition of any utterance, prior to nature and conventions, Plato begins to grapple with an imperative problem. Whether sophisticated or not, neither conventionalism nor the radical naturalism of Cratylus provided the criteria that would determine what is the correct, true, and suitable naming. So, if Cratylus – Plato's first teacher – was taken by the Heraclitian vision of nature as ceaseless flow and eternal impermanence, it was clear that the Platonic explanation, through Socrates' arguments in the dialogue, would have to grasp naming in a manner completely different from Heraclitus. Earlier in the interaction with Hermogenes and Cratylus, Plato had declared, "[T]here is a greater chance to find the correction of naming from unchanging essences" (*Cratylus*, 397 b). Names have an essential nature that does not change. Is that the kind of direct relationship that Cratylus proposed to be right? It would not be so since Plato's naturalism belongs to another order. Names do not imitate a world that is constantly changing. Names exist in relation to a world of unchanging essences. If it were the other way round, if names existed solely in relation to a mutant universe, knowledge – the basic condition for naming –would be inaccessible (*Cratylus*, 440 b-c).

Also crucial is to notice that Plato takes a stand against the supposition that human life was a direct and exclusive cause of divine intervention. His accent on knowledge transcends the type of divine knowledge that Protagoras claimed to be necessary. Plato is talking about the discovery of what is necessary and its moral implications; in other words, how the knowledge of necessity ought to make us act in a certain fashion. Thinkers before Plato called it Nature (Jaeger 1943: 219), but he transfigures this notion, not only giving it another name – *ananké* – but also conceiving it as a Higher Nature that would therefore transcend the transient and limited attributes of matter. That kind of nature is not the concession of inaccessibly divine forces. It is part of an intellectual discovery called philosophy. In such terms, the role of philosophical inquiry is to discover the profile of necessity. Necessity is the dominant trait in the world of forms that coordinate and shape the flow of experience. Its knowledge indicates what political, social, and moral rules to follow.

Plato's intermittent imagining of two worlds – one changing, the other eternal and suffering no mutation – does not put aside the possibility of error and falsity in naming. Because the two worlds are in a relation of imitation, and changes spring from timeless and necessary permanence, error is a recurrent feature of the sensible world. The translation of essences into existing things fosters mistakes. Imitation is always the loss of a pure and pristine world of essences.

2.7. The political sign

Ceaseless modification in the world of senses is an obstacle to finding truth. The type of knowledge to be reached if the object is in mutation has to be precarious, but that does not mean that it is utterly unachievable to assert anything about a changing object. Under this condition, it cannot be anything but a fleeting opinion.

What should be done to move away from the world of opinions? To escape from the predominance of viewpoints that cancel one another, it is necessary to find an objective foundation, from which it would be feasible to grasp the essence of things and ideas. Opinions are uselessly subjective. Based on subjective knowledge, one can only babble. In the dialogue *Theatetus* (147 b), Socrates professes that the naming of a thing has to presume knowing what it is. The consequence of all this was that the knowledge of essences should take precedence over anything else. With piercing irony, in Plato's early dialogues, Socrates points out recurrently that although his fellow-citizens were struggling to understand an abstract topic, all they would get were inconsistencies. Incapable of moving beyond the fluctuations of opinions, they did not have any inkling that to think is to transcend the web of appearances. They were still caught in partial explanations, filaments of opinions; they did not get that the right path to knowledge was to define what is essential in Virtue, Justice, or Beauty.

The viewpoint should be left behind. Socrates' critique of opinions is in direct opposition to what the sophists had been alleging and teaching to the Athenian youth. And of all such teachers the towering figure was still Protagoras. So much so that, in the *Cratylus*, Hermogenes admits that, in his confusion, he takes refuge behind the Athenian awe of Protagoras, though he does not agree with him at all. The crux of the dissension with the old sophist is whether "things have a permanent essence of their own" (*Cratylus*, 386 –ab). Socrates and Hermogenes then agree that naming and knowledge cannot be relative to an individual viewpoint. In that case, everything is reduced to a subjective appreciation, being thus irrelevant to take an objective standard and thus measure whether a man is good or bad. Socrates asks, *Have you ever been driven to admit that there was no such thing as a bad man?* Cratylus responded: No, indeed I have often had reason to think that there are very bad men, and a good many of them. *Well, have you ever found any very good ones?* Not many. *Still you have found them*? Yes. *And you would hold that the very good were the very wise, and the very evil very foolish? Would that be your view?* It would. *But if Protagoras is right, and the truth is that things were as they appear to anyone, how can some of us be wise and some of us foolish?* (*Cratylus*, 386 b–d). Truth is not appearance or viewpoint. If all viewpoints are valid, and if all appearances are correct, truth does not exist. What is true and good surpasses the instant; it goes well beyond the peculiar and idiosyncratic impression that it was true and good at that moment. Therefore, it is mandatory to reject the sophistic notion that what appears to be for an individual is the sole reality, and that the real world has no consistency of its own, being as varied as there are viewpoints. Truth is not to be suspended and put between brackets. Otherwise, the quest for truth would totally ebb like water on water.

Yet, Protagoras was correct in his conclusion that contradiction is a prevailing attribute of the world; but what was left unsaid was that contradiction abounded in a universe solely measured by opinions. Contradiction was an inevitable feature of the world of senses. Without an alternative to the subjective traps of the individual appearances and the viewpoints, the definite inference would be that there is no such thing as truth. Truth, however, is immutable and has to be safe from duplicity and multiplicity, which are symptoms of opinions and conflicting viewpoints. Situational ethics is grounded upon an antithetical conception of knowledge. Based strictly on the supremacy of the viewpoint, situational ethics can only be a twisted simulacrum.

To affirm that knowledge comes from the clash of opinions is false. In that case, without a normative standard, opinions cancel each other. Knowledge hovers above subjective evaluations. Take, for instance, the case of a sick person with fever (*Thaetetus* 178 b–179). Should one search for advice consulting a layman or ask a doctor what to do? Opinion and knowledge cannot be taken on an equal stand. One cannot be indifferent to whether the patient is treated according to the ignorant prescriptions of a layman or the knowledgeable recommendations of a doctor. On the other hand, a doctor cannot be right if the subject is shipbuilding. Socrates then refers to the case of Protagoras. If all viewpoints are equally valid, if all men are equally wise, why pay so dearly for Protagoras' classes if not convinced of his intellectual excellence and stature? Curiously, Protagoras publicized his thesis that there is no absolute norm; nonetheless, that was not valid for him, who charged for his instruction based on the assumption that he was wiser than the rest. As it is so frequent with Plato, his suppositions had far wider implications than what was immediately alleged. Plato's discussion on language went far beyond the question of language. At once, it disqualifies the wisdom of the political decisions of Athenian democratic assemblies. With acumen, Plato proves that a consensual political choice was many times purely the triumph of foolish ideas. Too easily, direct democratic systems allowed the irresistible ascension of imbeciles. It is to Plato's credit that he demonstrates that knowledge is a crucial condition for public choices. It is thus incumbent upon the citizen in a democracy to be informed and as knowledgeable as possible so that the decision is not infected with petty views and ignorant prejudices.

A political argument – any argument in fact – can be refuted not in relation to the number of people who approve of it. The majority can easily make a silly decision, or a destructive one, such as the Athenians had done when launching the Sicilian expedition. An argument can be rejected because of its own properties if it carries contradictions or unwarranted consequences. The logical form that identifies contradictions is this: if an argument P entails the consequence Q, then *non-*Q will necessarily imply *non-*P; if P implies both Q and *non-*Q, the argument is contradictory and thus false. An argument is invalid if the presented hypothesis entailed incompatibility and inconsistency of judgement. The delivery of the speech is less important than the logical content of the speech itself. No one is inevitably right; everyone can be wrong. The assembly's decisions are not immune from the same fate.

As his thought progresses, Plato, having abandoned his initial aim of recording the exemplary political action of the historical Socrates, transformed this exciting conclusion into an astonishing new turn of ideas. It is not anymore the insight that some could be right, and some wrong. The guidance of public affairs should not be at the mercy of unruly and unstable democratic assemblies. As Plato formulated consistently in both *The Republic* and *The Laws* (Pradeau 2004), true knowledge, and by extension who has it, should govern the life of the *polis*.

While the sophists were incapable of acknowledging errors as constitutive forces in knowing from which humans could not escape, for Plato the assumption became that all are always wrong. The crowd in the assembly was just a herd of ignorant citizens inebriated by their own false ideas.

The bitterness at the destruction of Athens through unwise democratic verdicts must have been a real shadow in Plato's mind. The decisive question now was, Who should govern this numbed crowd? The reply seemed foreseeable: if knowledge is the condition for reaching public decisions, who knows more should rule the city. Plato's reply and powerful argumentation presented in *The Republic* and *The Laws* inaugurated an extensive tradition of totalitarian justification of political ruling. Theocracies would deposit the power of social and moral decision on the religiously virtuous, transforming religious leaders into political rulers. Rousseau would see the situation under a similar, although secular light. Religion should play no role in political decisions; the nucleus of the decision was the collective will (*la volonté collective*) capable of exercising cohesive force over the individuals. Karl Marx would write and demonstrate that, in an industrial society, the ruler ought to be the oppressed economic majority: the proletariat. The question that Plato tried to answer – "What is the natural ruler?" – displays the same logical form as "What is justice?" or "What is the good order?" The provider of the answer to the definition of justice and good order ought to be the enforcer or the ruler.

Mainly, that was the shedding away of the initial stand of early Platonic dialogues, describing a critical Socrates in action. In them, the question was not "Who governs?" but "How should the city be governed?" The reply was through the falsification and criticism of any assumption. Protagoras' emphasis on strong speech as the expression of communally held opinions was mere conservative convenience, morally inadequate as well as a political disaster. Much more important was the willingness and the courage to scrutinize closely any proposition, subjecting it to the strongest assail.

The initial criticism of Plato's early Socratic dialogues was subdued and transfigured into the justification of non-democratic practices. As the individual citizens should be, signs – names, words, but not only that – are the effects of rules that conduct and coordinate their production and circulation in social life. Rules are the mental projection of essential forms. Rules preceding the actual performance of users establish their correct and adequate application. Individual users are coordinated by and submitted to rules. Although it may have been an unnoticed and unintended after-effect of Plato's domineering vision of signs, politics, and society, Roland Barthes (1995: 803) was painfully and Platonically accurate when he declared that language (*la langue*) was fascist. The adjective *fascist* is, however, better applied to post-Platonic conceptions of language, rather than to language itself.

Barthes's provocative sentence, pronounced in 1978 during his inaugural lecture at the Collège de France, owed its thrust to the structural notion of code, but it is also true that his statement revealed the long-standing and mostly unspoken influence of Plato over the canonical conceptions of language and sign production. It is not merely that social legislation and cultural bonds are imparted through the circulation of linguistic signs. It is too simplistic to state that the rules of sign production are acts of political power. Rather, the idea is vaster than that: signs impose a classification over the natural world, separating humans from other forms of natural life. Rules coordinate the production of messages, determining specific conditions of sign usage. As in a totalitarian state, as in Plato's social utopias, it is not merely a brutalizing censorship. It is an obligation: rules compel what to say. The narrow focus on rules generates a distorted view of sign production. Sign usage is not only the actualisation of prior rules. When signs are used, one can observe the creative transformation of their formative conditions. Nothing is more anti-Platonic than that.

The recognition of a world of permanencies – in itself the cause and the condition for flux and relentless change – brings with it a series of connected consequences, some of them metaphysical, some moral, some political, and some of them semiotic, all closely related to Plato's prescription to the health of the *polis*. The consequences are manifold. They presume a distinction between what is essential and what is accidental. Moreover, from this distinction it should never be inferred that essences and accidents live side by side. It is mandatory to concede their asymmetric hierarchy: essences coordinate accidents. Furthermore, such hierarchy entails a relationship of subordination: essences rule accidents. That is more than cosmological. Akin to the traditional Greek fascination with the overlapping of *cosmos* and *polis*, it is fundamentally a political recipe; otherwise, the good order – the uncorrupted and ideal *eunomia* – would never exist. What is mutant and circumstantial is not to be confused with true and fixed norms. Rules have to be kept intact. They embody what should be adopted as collective knowledge and are irreducible to any individual dimension. The knowledge of basic and eternal forms structures matter and individual performance in society, subjugating the singular to the plural. Fixed forms and crystallized rules are forces of permanence coordinating and organizing a frightening world in perpetual fluctuation. Although not frequently recognized as a by-product of Platonism, and many times disguised under the form of an unabashed secularism, the canonical notion of culture as a system of coordinating rules acting upon the will and the behaviour of individuals is a tributary of the lasting legacy of Plato's ideas. Several centuries later, Augustine of Hippo will give it its definite shape.

Chapter 3
Contract

When Aurelius Augustinus, no longer a catechumen, having been baptized by Ambrose, bishop of Milan, at the Easter vigil of 24-25 April 387, arrived at the left bank of the mouth of the Tiber River with a group of travelling companions, they knew immediately that it would be impossible to reach Carthage any time soon. During the summer of 387, Ostia, a Roman seaport that had been both the defensive bulwark of the Empire and the route for feeding the Imperial capital, was under naval blockade. Augustine, his mother Monnica, his son Adeodatus, his brother Navigius, his friends Alypius and Evodius would have to wait for the outcome of the struggle between two Catholic emperors disputing over the Roman Empire. Augustine's group lingered there, not lodged temporarily in a local inn, but hosted by friends and acquaintances, "probably senatorials with African connexions" (Meiggs 1960: 213), for approximately one year and a half. Without reaching North Africa, Monnica would die, perhaps of malaria, and be buried in Ostia. Words praising her chastity, righteousness, and devotion were chiselled on her tombstone.[1]

Magnus Maximus, the usurper, blocked the seaport hoping to bring down Theodosius. Augustine had almost nothing to say about the blockade and other historical events. Though selection is an inevitable feature of literary composition, the choice of what is to be left out is as significant as what was written.[2] So why is nothing mentioned of the siege of Ostia, not even the relief Augustine must have felt when, after Theodosius' victory, the blockade was over and the group could then move on to North Africa?[3] In Book 9 of *The Confessions* Augustine admitted that he had kept quiet about many things of his life because he was writing in haste.

The excuse, however, is unpersuasive. At this point (*conf.* 9, viii, 17), Augustine's not easily classifiable masterpiece was nowhere near its end. Augustine proceeded to write four more books integrated into *The Confessions*, all of them theoretical in nature, with scant, indirect allusion to his personal life: one dealing with memory, two others about the Christian conception of time and creation, and the last on the peace to be found in the resurrection of the souls reposing in their meeting with God. After describing external events of his life, Augustine wrote the complementary section on memory, reflecting on the inner dimension of the individual, without which conversion would be impossible, and then he came to the final three books dedicated to the Scriptures, the canon that wrought his Christian identity. Among other commentators, Peter Brown (1993: xiii), Augustine's leading contemporary biographer, recognized the jolt separating the events of the first nine books of *The Confessions* and the four remaining ones. Chronologically, the gap between the initial biographical narrative and the last books was more than one decade. The first nine books had an dominant mood of recollection, a re-enactment of selected past events of Augustine's life of thirty-three years, drafted not to leave a vainglorious record of his life, but to offer a testimony of God's glory and mercy toward a repentant sinner. To accuse and to expose one's errors were worthy modes of praising God.

Incessant anguish and shuddering anxiety pervade *The Confessions* until the end of the ninth book; then the writing briskly changes: the last four books seem to have been composed in the solitude of Augustine's study, products of profound meditation over the text of the Scriptures by the man who was now, in 397, the Catholic Bishop of Hippo. Much later, written sometime around the period of 426 and 427, indeed three years before his death, *Retractiones*, Augustine's reconsideration of his long and prolific literary and doctrinaire life, made little or no adjustments to his most popular work among the brethren (*ret.* 2, 32, 1). Augustine mentioned the awkward way he described his feelings after the death of an anonymous friend, censored his own obscurity about what the called "the firmament," but nothing else is revised: no reconsideration is given to anything outside of his inner life or his doctrinaire action.

Augustine's lack of interest in historical happenings could well be the natural indifference of a Christian renouncer, wholly wrapped up in his longing for a place to establish a monastic and ascetic community. Why would he be remotely interested in the struggle for earthly power and glory? In fact, Augustine had firmly announced in a work written in Cassiciacum, a retreat probably near Milan where he went with his mother and a group of friends before reaching Ostia, that, since his early manhood, he wanted to acquire wisdom (*sapientia*), but of a different kind than the type Cicero had advertised in *Hortensius*. Augustine yearned for wisdom beyond worldly success – not *ambitio saeculi*, but a wisdom that could calm his unquiet heart – a wisdom grounded on beatitude, as is hinted in the opening lines of his *Confessionum libri tredicim*: "our hearts are restless till they rest in You" (*conf.* I, i, 1). Our destiny in this earth is to burn restlessly: the only peace is the peace of God. In the *Soliloquies* dictated in Cassiciacum, Augustine argued with a dramatic entity he called Reason, expressed as a sudden voice, erupting into his conscience. Reason asks what Augustine would wish to know: "I want to know God and the soul. Nothing more? Reason asks. Nothing more," Augustine replies, "nothing whatsoever" (*sol.* I, 2, 7).

3.1. Thorns in Augustine

If perfect happiness lies not in the mere search for truth, but in knowing it, how can one know that one knows? Augustine tried incessantly to find a way to wisdom in the years prior to his conversion. With that taken into account, the first nine books of *The Confessions* could never be seen as the simple narrative record of a conversion. In Augustine, there is no room for adopting Christianity as an effect of credulity. Augustine's insistence on the role of faith (*fides*) in his conversion points to a grander design. His espousal of the Christian faith is an altogether distinct choice from mere obedience and surrender to the religious habits of his uneducated, pious mother, always deeply immersed in the liturgy of the Catholic Church and constantly praying for the conversion of her son. That was not enough. Habits cannot be the motive for such an important decision that would change one's life. Conversion must be a radical and revolutionary transformation, the outcome of intense thought and dedication, capable of transcending finally the inevitable conformity to previous habits. The road to God begins with the elimination of habits: "The law of sin is the violence of habit by which even the unwilling mind is dragged down and held, as it deserves to be, since by its own choice, slipped into the habit" (*conf.* 8, v, 12). Conversion must necessarily be a decisive act of will, an act of personal liberation; and, as such, it is incompatible with whatever weakens the resolve to reinvent oneself: for most people, habits pave the path to perdition, not to redemption.

When a young Manichean in North Africa, Augustine heard and may have agreed with the claim that habits difficult or impossible to break are evidence that conflicting forces are constantly at work in life, always shaping human destinies. The world is a stage for endless struggle. Sometimes the principle of light and goodness prevails; sometimes its opposite, the principle of darkness and evil, triumphs. The Manichean argument seemed to Augustine a sufficient explanation for a universal trait of life; furthermore, it had the appearance of a purely rational argument for the persistence of evil in this world. The discovery of and the adherence to such an elaborate explanation should have been flattering to Augustine in his early youth, "just when he was beginning to feel the confidence of a very young man in his own intellectual powers" (O'Meara 1965: 69). Augustine's Manichean conversion is also closely related to his disenchantment after reading the Scriptures. It is obvious that Augustine's disappointment in the Scriptures went far beyond the dislike of their flat, inelegant eloquence. Augustine recognized several inconsistencies in the Biblical narrative. One of his sermons – *Sermon 51* – reviewed the contradictions in what was supposed to be the word of God. For many young North African intellectuals, Manichaeism appeared to be the proper alternative to the utterly humble propositions of Christianity. Manichaeism shared visible commonalities with Christianity, but the differences were striking: it dismissed the Old Testament; Genesis was mere forgery; the prophets and Biblical characters displayed intolerable vices; and yet Manicheans never rejected Christ, they considered Christ central to their doctrine, in conjunction with the Apostle Paul. Manichaeism was a Pauline heresy, presenting itself as a kind of pure, proud, and rational Christianity. Manicheans claimed to be members of the true Christian Church (*ret.* 1, 6, 2).

Augustine realized gradually that the Manichean doctrine was sheer fantasy. Its rational explanation was untenable, and its mythical narratives about the origin and the creation of the universe were a laughable alternative to the Genesis. As a whole, the Manichean cosmology was simply arbitrary and implausible: nothing but falsehood and nonsense. Why Augustine would remain faithful for nine years to a doctrine that he came to despise and oppose so systematically? The attraction had to be the strong Manichean reply to what seemed to him a fundamental and mysterious question: "Whence comes Evil?" (*conf.* 3, viii, 1). Ironically, in his later years, adversaries accused Augustine of being a Manichean in disguise, because all his life he had presented recurrent answers to the same Manichean question on the nature of Evil. Augustine's reaction against one of most ferocious accusers, Julian, Bishop of Eclanum, was passionate: "If this mean nothing to you, behold, I condemn and anathematize what you say the Paternians and Venustian hold; and the Manicheans, I execrate, I condemn, I detest them together with all the other heretics" (*c. Iul.* 5, 7, 26). Augustine asked Julian to compare what he had been writing all along with the opinion of Manicheans: "I not only detest and condemn in faith and word the mixture of two natures, one good, one evil, from which spring their whole imaginary raving, but I also oppose it by resisting you, their supporter" (*c. Iul.* 6, 21, 66).

Augustine's counterattack calling Julian of Eclanum a Manichean was unreasonable, but his outrage was, without a doubt, sincere. His past, however, was returning to haunt him: when among Manicheans, Augustine found satisfactory their reply to the puzzle of how the Creator could be responsible for the existence of Evil in this world. Either God created everything, and that meant the creation of Evil, which contradicts the goodness of His nature, or there is an autonomous substance that God did not create. God could not have created Evil without sacrificing His inherent goodness. Augustine's early fascination with Manichaeism betrays a staunch fidelity to a definition of God as a Perfect Being that – it must be said – he always held dear. His consistent quest for God is patent: "since the day I learnt of you, I have never forgotten you" (*conf.* 10, xxiv, 35). Augustine never declared that he suddenly came to God; but, rather, that he was lost, astray along misleading avenues and God found him.

For any individual who valued rationality, the Manichean answer to the puzzling existence of Evil was intellectually unsatisfactory. It denied the essential attribute of God, that of being the Creator. Manicheans never explained how the Omnipotent God would have to be involved, on equal terms, with a created substance; and, what is worse, lose the struggle. That would be totally against the notion of a potent Creator, a Creator that could only be halted by the rainbow of His will.

Although initially unclear to Augustine, and not formulated exactly in this way, the Manichean answer to the question "what is Evil?" was just fuzzy thinking. If any question takes the form of "what is *X*?," it generates the temptation to find a reply through the direct indication of an entity, a substance, or a being; but frequently, the temptation yields no more than a linguistic mirage. A more complex examination of Evil succumbs to a bewildering variety of aspects, irreducible to the identification of one single principle. Evil can be metaphysical (*malum,*), logical (*falsitas*), psychological (*perversitas*) or moral (*iniquitas*): from these prismatic reflections a cohesive profile is drawn. Such complexity completely disappears in the Manichean doctrine. The Manichean reply was a gross simplification. Moving from the question "what is *X*?" to the answer "*X* is __," the Manicheans concluded that Evil is matter.

The denial of the goodness of matter was more than just a logical conclusion; Augustine must have heard, during his early years, a stream of narrative and tales of exemplary ascetics who had renounced the material world as the condition to access God. Contemporary philosophers would simultaneously assert that the spirit is not only altogether different, but also superior to matter. In this mental ambience, the Manichean reply seemed right and appropriate to someone, in his own words, "terrified by superstition" (*util. cred.* 1, 2) Manichaeism rationalized asceticism; but if one considered that the reply begged the question, it was clear that Evil could not belong to the order of being. Evil was precisely its contrary: Evil was non-Being; therefore, it could not be at the same level of God, the Supreme Creator. Under the scrutiny of reason, the superficial rationality of Manichaeism crumbled apart. Reason seemed to be both the method and the standard of judgement to search for God; and yet at that point it did nothing but falsify logical errors and, negatively, raise doubts: where would the positive knowledge (*scientia*) of God be found?

3.2. The insufficiency of rhetoric

Augustine must have felt that to look at his past was useless. In Augustine's time, the available education was "frankly pagan" (Brown 1967: 36). His father, Patricius, a taxcollector (*curialis*), had to scrape resources and find a sponsor, a wealthy relative, Romanianus, to pay for Augustine's costly training (*c. acad.* 2, 2, 3). In the schools, a young boy would receive a meagre education. No attention was given to the inquiry of the theoretical foundations of what had been taught. Augustine would be asked to memorize a handful of classical authors: Virgil, Cicero, Sallust, and Terence. At that time, Roman education followed a trend that went back to the tradition of Gorgias of Leontini (485–380 B.C.E.) and Isocrates of Athens (436–338 B.C.E.), with great emphasis on the emulation of models. One of the main motives for the Roman adoption of Greek culture was that both societies placed an extremely high value in the mastery of rhetoric: Greek and Roman political systems required that an active politician "had to know how to get on the right side of the crowd, how to influence the voting in the assembly, how to inspire the troops, how to harangue a tribunal" (Marrou 1956: 328).

Education was then reduced to the development of a skill to be used in the appropriate circumstance. The student would memorize the acknowledged masters of eloquence to such a degree that he would become impregnated with the chosen models, and then the effect would be his transformation into an improved orator. Nothing else mattered. Augustine would refer to his education as well as to his initial teaching career with the bitterest resentment. It seemed unpardonable to him the utter disregard for fundamental moral concerns: What is the purpose of a speech? Is it the expression of the truth, or just mendacity, so many times mere plausibility clothed in verbal eloquence to cover its malicious goals? Recollecting the pangs of his education, Augustine accused his elders: "They gave no consideration to the use that I might make of the things they forced me to learn; their objective was merely to satisfy the appetite for wealth and glory" (*conf.* 1, xii, 19). Although the training was sclerotic, fixed, and rigid (Marrou 1938: 51), the social rewards were attractive for a family of Augustine's means and background. The schools had become the training ground for the Empire's administration. Several examples were around to encourage that hope. The Emperor Domitian began in the first century the tradition of honouring professors with political posts when he conferred consular rank to the rhetor Quintillian. Ausonious was also made consulate and prefect of the praetorium of the Gauls, based exclusively on his verbal eloquence: Ausonious had no prior administrative experience. Themistus was appointed senator and archonproconsul of Constantinople as recognition of his excellence as a teacher (Marrou 1956: 412).

Augustine's success could come from his eloquence in the forum or in the classroom, but that would mean the acceptance of what he felt a distraction from the ambitious goal of finding happiness through the knowledge of God. Augustine's rejection of this kind of future was expressed, in a mixture of relief and indictment, when he recalled his resignation from a teaching post in the Imperial educational system, just before his definite conversion to Christianity. Augustine's words were rash and unambiguous: his previous secular activity filled him with disgust (*conf.* 8, i, 2). At the same time, the example of Marius Victorinus, a rhetor also of high learning and culture, a teacher and translator of Prophyry and Plotinus, was encouraging. Augustine had read Marius Vitorinus' translations of "some Platonic books," when Simplicianus, an aging Catholic bishop, closely connected to Ambrose, described to the younger rhetor Victorinus' passage from intellectual acceptance to becoming a baptized Christian (*conf* 8, ii, 3).

A wide wave pushing in the direction of Augustine's ultimate option for Christianity was gathering momentum. He had already come to the conclusion that worldly ambitions and verbal eloquence, an indissoluble blend in his and so many of his contemporary minds, would have to be shed away if the goal was happiness and beatitude. Besides lavishing praise on unworthy and powerful officials, as well as producing verbal delight in admiring and gaping audiences, rhetoric could only spread doubts. Conversion, though, ought to occur when the search for certitude is reached. Truth-likeness (*vere similia*), the final and nonetheless limited goal of persuasive speeches blatantly unconcerned with Truth, was unsatisfactory: the criterion of verisimilitude was positively harmful because it could not bar the invasion of doubtful arguments into the converted mind. Augustine remembered how, when a brash Manichean, he had disputed with rhetorically unskilled Christians, having always crushed his opponents. Those were sour triumphs, for he realized later that he had not defended what he now deemed to be the truth. When the debates were over, his victories filled him with pride and arrogance (Brown 1967: 48). Augustine's *Confessions* thus constantly berated rhetoric as loquacity (*conf.* 1, vi, 4), as a vacuous exercise in futile conceit, or as mere mendacity (*conf.* 8, ii, 4). Under the sway of eloquent speeches, "the soul is thus carried about and turned, twisted, and twisted back again. The light is obscured from it by a cloud; the truth is not perceived" (*conf.* 4, xix, 23). Christian conversion should follow another trail: "the most peaceful and proper way to seek the truth with God's help would be by questioning and answering myself" (*sol.* II, 7, 14). No external speech could ever account for the inner individual experience of the converted mind.

3.3. The importance of wisdom and happiness

In a letter (*ep.* 118) written around 410 or 411 to Dioscorus, a young admirer of Cicero and classic ancient rhetors, Augustine tried to answer the question of whether or not the pagan study of rhetoric was worthy. For Augustine, once again, the pursuit of vain ambition and the training in rhetoric were inseparable. Because of this, his initial reaction was to ask Dioscorus: What was the purpose of "all this too ardent study, useless to you and troublesome to me?" Much more urgent was the devotion to moral questions that lead to the transformation of one's destiny. If one abandoned the goal of personal reform, nothing would be left but the display of pointless eloquence. In his long reply, Augustine could not hide his impatience with Dioscorus' previous insistence to have answers to such questions; Augustine felt that replying was a colossal waste of time, which diverted and subtracted from his many cares and obligations as the Bishop of Hippo. Verbal embellishment and excessive eloquence, he had been preaching repeatedly, were not the primary goals of Christian preaching. Dioscorus should not be ashamed of ignoring the finer points of rhetorical theory. If he dedicated himself to the study of pagan rhetoric, he would only get "the hoary and worn-out falsehoods of so many philosophers." The temptation ought to be avoided; though, as he admitted, earlier in his life Augustine had once sold such wares to little boys. It had taken so much time for Augustine to find anchor and peace in the warm embrace of the Church: Why should Dioscurus repeat his errors? The tradition and the objective of philosophical discussion were foreign to the Christian pursuit of God, certitude, and truth. At that historical moment, a Christian ought not to get into endless discursive entanglement of pagan philosophical enquiry but to face the schisms around them. More important was "to think of the heretics, who call themselves Christians, rather than Anaxagoras and Democritus."

According to Augustine, the Christian quest had to be always for the unalterable truth; therefore, it could not be dictated by the shift of philosophical opinions. He was sure that the truth would shine anyway, regardless of any debate with its detractors. Thus no help would come from the familiarity with pagan philosophical schools, whether to criticize the Stoics and their puerile identification of wisdom and apathy, or to attack the materialistic Epicureans who searched for happiness in the transient satisfaction of bodily desires, much less to assail the Platonists for their narrow quest for truth following the narrow although ascending rationality of the soul. The problem with all pagan philosophical schools was their disregard for God.

If called a rhetorician or philosopher, Augustine would be affronted. Rhetoric should be criticized for the harm it did, whereas philosophy, without religious concerns, should be written off for what it could not do. It was pathetic that a rhetor would frequently choose to deliver an argument that his audience would find plausible rather than to teach what is true. Catering to the whims of audiences was nothing but abject servitude. Also shocking was how the training of rhetoricians would blindly advise that immoral option of preferring persuasion to truth without the slightest qualm. The consistent disregard for the truth amounted to the dissemination of mistakes. Mistakes are incompatible with justice. The sacrifice of truth for worldly success and public acceptance was another indication of the deep moral flaw in everything that concerned rhetoric. Rhetors would shamelessly and irresponsibly trumpet a malicious proposition: a plausible falsehood can be more valuable than an unlikely truth (Fortin 1975: 89). If that was so, mendacity and persuasion were so intertwined that it was impossible to discriminate them.

During his years of teaching, Augustine wondered: What was the point of an empty life devoted to the delivery of subservient praise of the powerful? To consider the mission of an orator as having to simultaneously please, move, and instruct was in the end sheer hypocrisy. To please, to move, and to instruct had different aims. How could one instruct without the compass and the commitment to finding the truth? Truth could be unpleasant and unmoving. If having to choose between pleasing, moving, and instructing, what should an orator do? With his training as guidance, the orator could be easily driven not to prefer the truth but to shamelessly spread falsities, disguised as probabilities.

Before Augustine, Cicero objected to the restricted training of orators in verbal display. Cicero's *De Oratore* (2001 [circa 55 B.C.E.]) was much more than a handbook of rhetoric. Cicero was advocating radical changes in the schooling of orators, in practical terms a complete reform of the educational system. Besides backing up the usual requests for becoming a good orator, whether underscoring natural or acquired abilities, such as developing a potent voice, refining elegant gestures, or perfecting a verbal style, Cicero argued that the orator would need to ground his delivery on solid knowledge. The deliverer of speeches should master Roman laws, as well other modes of knowledge that would grant him the capacity to answer fundamental and universal questions about what makes human beings. The subjects of the orator's knowledge should be history, jurisprudence, and philosophy. For that reason, rhetoric became a subordinate part of the universal program of improving citizens. Education should train individuals who would have the task of being *politior humanitas* (Gwynn 1966 [1922]: 102), of improving humankind. The ancient educational model should be reformed; so far, Cicero claimed, it had been merely *puerilis institutio*, the training of the young and the immature.

Overall, Augustine clearly read too much into Cicero's text when he ran into the *Hortensius*. Cicero's emphasis on wisdom was the anticipation of a clue to mark out his personal rupture with the pagan culture that had shaped him up so far. And yet, that was not everything: Augustine's claim in the dialogues he wrote just after his baptism – that the liberal arts could provide the ground for the discovery of God – indicated that he may have felt that Christianity could be a new horizon not only for the uneducated and the humble, but also for a growing class of cultivated people who were more and more sympathetic to the Christian way of life. The words with which the Bishop of Hippo described, in his *Confessions*, how he came to pursue wisdom indicated that Cicero was no more than a starting point. Reading "the usual curriculum, I had already come across a book by a certain Cicero, whose language (but not his heart) almost everyone admires" (*conf.* 3, iv, 7). Here, Augustine's tone is not one of full acceptance, though not one of complete rejection either. Since then, interpreters have discussed extensively the meaning of Augustine's words "*cuiusdam Ciceronis,*" a certain Cicero.[4] Did he look down on Cicero? Was it just a normal expression, not really pejorative, or was it like O'Donnell (1992b: 164) observed: "the expression signifies the vanity of a fame like that of Cicero in the presence of God"? Surely, the word "*cuiusdam*" qualifies Cicero as a stranger to a Christian life. Nonetheless, Cicero's zeal for wisdom had been inspiring; it changed Augustine's feelings and prayers, although a pagan like Cicero would have never achieved *sapientia*. For Cicero, the search could only lead to failure: he did not have the knowledge of God as his dominant goal. Augustine insisted that it was not sufficient just to pursue wisdom. Wisdom must be reached and ought to provide *the right way of life that leads to truth* (*c. acad.* 1, 5, 14).

3.4. The demise of the classical tradition

Augustine obviously recognized that his idea of wisdom differed considerably from Cicero's; for the future Catholic bishop, the ultimate homeland (*patria*), "the soul's proper habitation" (*quant. an.* 1, 2), ought to be God, not at all the pagan conception of the divine, shattered by the belief in multiple gods. He had yet no idea where to find Him, but inspired by the idea of a sole God, Who created all things in His likeness, Augustine guessed that the search would be always for unity, whether between thinking and existing, between talking and doing, between Creator and creature, or between will and deed. That was Augustine's aspiration: he understood that in God there is no duality. Later, his attraction to the Manichean heresy would serve as undeniable evidence that the uncritical acceptance of duality fostered regretful errors. At least, compared to other influences coming from his pagan education, Cicero could somewhat suggest to Augustine the virtue of unity. Cicero had indeed indicated that the rhetorician should transcend factions and divisions. Thus, Augustine took delight in Cicero's exhortation "not to study a particular sect, but to love and pursue and hold fast and strongly embrace wisdom itself when found" (*conf* 3, iv, 8). However, Cicero was really talking about another kind of wisdom, the wisdom of tolerance and respect for the variety of viewpoints in life, the wisdom that is the opposite of the unreachable absolute certitude. In his own way, Augustine also rejected arrogance and pride, and yet he craved certainty. His justification would be that our mind cannot be understood, even by itself, because it is made in God's image (*Sermon* 398). *Sapientia* is not an aimless quest: it has to be guided toward God. *Amor Dei* "cannot be an uncertain feeling, but a matter of conscious certainty" (*conf.* 10, vi, 8). If in God there is no uncertainty, to know Him would necessarily dispel doubts. Here, Augustine took sides against the scepticism that lured Cicero in his dialogue *Academica*.

As the reverse of certitude, probability is nothing but imperfect knowledge. Probability does not even allow the detection of an error; so, to be content with probability is to be satisfied with remaining within the confines of narrow human knowledge and therefore never aspire to true and definite wisdom. Any human being can make predictions that are eventually fulfilled. Because of God's omnipotence, probable and occasional certitude are allowed even to the malignant, the miserable, and the wretched, whether humans or demons, whose kind of knowledge is without the charity and the love of God; but it is knowledge anyway. If right guessing is granted even to nefarious beings, how can one totally dismiss certitude? In effect, Augustine would find it quite normal that an Egyptian pagan god, Serapis, could foresee the destruction of his temple (*divin. daem.* 7. 11). Altogether different would be to "foresee the changes of the temporal order in the eternal and unchanging laws of God, which live eternally in his wisdom, and, by participation in the Spirit of God, to know the will of God, which is supremely certain and supremely powerful." (*civ.* 9, 22). Augustine could dream of such heights: he acknowledged that humanity was made in the likeness of God. If the mind of even wicked individuals can attain occasionally right guesses, why should one be content with a merely probable answer to the mysteries of life?

Augustine felt that doubts were inevitable in the mobility of thought groping for truth; but that did not sanction the sheer folly of doubting everything (*trin.* 15, 12). Systematic scepticism denied any viable repose in a definite truth. This kind of radical scepticism favoured not occasional doubt, in this case an obstacle to overcome, but hyperbolic mistrust. It amounted to the absolute denial that true knowledge can be achieved. Scepticism thrived on impossibilities, the impossibility of knowing what is true, what can be true, as well as what is not true at all. At best, the alternative was to put everything between brackets and to resign oneself to truth-likeness: Augustine could never find that tolerable. How would it be possible to have truth-likeness without reference to truth? Despite Cicero's request to reform the nature of rhetoric, nothing had been altered. Augustine also repudiated Cicero's statement that "a man is happy if he seeks the truth, even though he should never be able to find it" (*c. acad.* 3, 7); his expectation was of a different kind: happiness was actually finding the object of the pursuit.

In October of 384, probably through his Manichean connections, Augustine presented an oration to Symmacus, the prefect of Rome, a man of literary talent (Brown 1967: 66). Augustine then won Symmacus' recommendation to a teaching position in Milan.[5] Having rationally found irreparable contradictions in the Manichean doctrine and yet incapable of accepting the Christian Scriptures, which paled in comparison to the eloquence of Cicero, Augustine adopted, although half-heartedly, the scepticism of the Academics. Wherever he turned, something was missing. Reason guided him to an oppressive labyrinth of doubts, to a point that his cherished ideal of discovering wisdom seemed unattainable. Augustine found himself in the fleeting state of total suspension of judgement (*conf.* 5, xiv, 25). Philosophy promised love of wisdom, but offered no consolation to his growing anguish. What he encountered in translations or introductions to philosophical concepts, probably through his readings of Cicero, made no reference to Christ, whose name Augustine understood, since his childhood days, to be the route to salvation. He became then a catechumen in the Church, simply because he had been raised in a Catholic family. Augustine's definite conversion required clearer light. If Christianity was ever to be a fully satisfactory choice, radical scepticism had to be definitely defeated.

3.5. Undoing a labyrinth of doubts

Once more, Augustine met the limits of rational investigation. Even if sometimes declaring his enthusiasm for theoretical speculation, the tools of philosophical inquiry were inappropriate for what he desired. The contentions of the major philosophical trends of his time were secular. The Epicurean accent on sensations was not only against Augustine's ascetic nature, but the privilege given to the unfettered satisfaction of desires seemed to be in contradiction with the very idea of morality; the ethics of Stoicism were attractive to Augustine; but its radical materialism also repelled him; the hyperbolic scepticism of the *New Academy* was useful only in furnishing rational arguments to debunk both the Epicureans and the Stoics, but scepticism was entirely at odds with his desire to unearth the blissful knowledge of God. It could not have been difficult for Augustine to understand, at this point, that in the world of arguments he would not come across the type of felicity that he longed for so eagerly. The ground of his choice for Christianity would have to come from his innermost experience.

As *Confessions* unfold, Augustine's narrative records a journey of tense wish to find signs that would unravel the veil of doubt and anguish. His intellectual background would not allow him to rule out what he read in books as potential signposts to beatitude. Indeed, the reading of unspecified Neo-Platonic texts gave what he believed to be both the confidence and the method to arrive at the aspired unity, a full merging with the Plotinian *One*, which he understood to be God. Plotinus and his commentator Porphyry made him aware that philosophy and religion were not adjacent fields, but enclosures within the same territory.

The initial step toward beatitude should start with the mind willing to reach the highest level of being: and the final goal would be attained when the inner mind felt an overpowering sense of happiness only possible in the union with the Creator. At this moment, all duality and dissimilarity between creature and Creator would have vanished. Augustine hoped that the soul's eyes would experience "not the light of the day, obvious to anyone, nor a larger version of the same kind which would, as it were, have given out a much brighter light and filled everything with its magnitude" (*conf.* 6, x, 16). The feeling would be ineffable and beyond description: it would not be the natural light, but a radically different encounter, irreducible to all kinds of light.

For that to happen, the seeker should be moved by the intellectual admission that there is immutable truth, of a noncorporeal nature, belonging to a world that is beyond human intelligence. The seeker should then proceed by degrees, shedding away what participated exclusively in the flawed world of the senses, abandoning matter as the journey progressed, ascending slowly toward unity, leaving behind "the region of dissimilarity," where any existing thing is because it is made in God's image and is not, because it is not God. In contrast to the perishing material world, God is absolute similarity, resting in the perfect consonance of His Will without change or dissimilarity. Although contradictorily immersed in a region of dissimilarity, sunk into a hopeless blend of being and nothingness, the created universe bears likeness to the Creator. The seeker of God should thus soar above multiplicity and end up grasping totality, simultaneously experiencing the transcendence of finitude and matter and finally achieving happiness.[6]

Following Neo-Platonic recipes, Augustine tried to find beatitude, but his attempts through introspection failed, perhaps not as the direct result of the method, but because of Augustine's guilt: he felt that he was not sufficiently pure (Courcelle 1968: 166). Augustine's early failure in reaching beatitude through sheer and straightforward will indicated that human intention alone was simply not sufficient to accomplish his task. He had a glimpse of the ascension, but the insight was, nonetheless, unstable. In an admission of guilt, Augustine wrote: "I was caught up to you by your beauty and quickly torn away from you by my weight. With a groan I crashed into inferior things" (*conf.* 7, xvi, 22). The sense of loss was devastating; he had not reformed himself and had to return to the lower regions of being, pulled by the remembrance of his sexual habits.

The total experience of illumination was proscribed to him because he rationally wanted what indeed he did not have. Augustine eventually came to realize that illumination is reached when reason is not a light to itself, but when it shines participating in the light of another (*civ.* 10, 3). God was Augustine's missing Other. Unwilling to renounce sex, Augustine was at this moment hopelessly caught in the soil and the snares of the secular world. For him, the explanation about what had happened was found in the Scriptures: the body is corruptible, and weighs down the soul (Wisd. 9: 15) Again, a text – this time a Scripture – furnished the key to him. Books and his reaction to them continuously signalled the whole trail leading to Augustine's conversion.

Augustine's mode of reading and finding sense in books was akin to his method of personal inquiry. Whatever sign came to him, it was never an immanent representation. Even in the natural world, in the world's stage of *signa naturalia*, a sign may seem to have just an intrinsic content; but signs point to what is not the sign itself.

Natural signs are, however, impoverished representations; they have no intention, will, or desire. While natural signs indicate potential knowledge, they themselves are not cognition. Cognition and will are effects of active minds, free from the strictures and limitations of matter. And yet even without this thick layer of intended meaning, the interpreter can always think of something beyond the impression the thing itself makes upon the senses (*doctr. chr.* 2, 1, 1). As the natural world is saturated with bewildering and contradictory signs, the inner mind sorts them out; but caught in the error and the corruption of the physical nature, even the most capable of interpreters can make serious blunders, as it occurred when, by the performance of miracles, pagan gods persuaded even intelligent men like Porphyry either to offer sacrifices or surrender to the irrational worship of demonic entities (*civ.* 10, 18). That happened because the pagan miracles were taken as immanent signs. With some thought, Porphyry would have realized how wrong the pagan religions were. Paganism did not see that God ought to have made human beings in His likeness; it was puerile to conceive of a plurality of gods. Multiple gods are a human fabrication. The pagan assumption was illogical; it contradicted the hierarchy of the created universe, submitting the Divine Maker to lower creatures. Because of this misconception, pagan gods were infused with appalling moral, emotional, and intellectual disturbances.

3.6. Among signs

While the visible signs conveying the sense of performed miracles did not change, it was undeniable that the same sign could mutate into incompatible and incongruous meanings. The pagan evaluation led to the worship of their gods as true divinities, but from the Christian perspective, such devotion was just sacrilege, nothing but the reverence of false idols. How could the same sign carry conflicting meanings? That occurred because the same sign was understood in relation to opposing abstract sets of values and beliefs that transformed the apparently unchanging configuration of the sign. One interpretation excluded the other; they could not be equally valid and entail the same meaning. The inference was that the signs in themselves were mere vehicles to be understood in relation to an array of assumptions that underlined and constituted them. Augustine argued that the pagan error leading to demonic worship could be avoided by going beyond the signs; whoever upheld the Christian doctrine would clearly see the evils of paganism.

Invisible elements at work in a sign do not make it less real. Indeed, it is quite the opposite: what is invisible grants reality to signs. When a sign is conveyed to another mind, its meaning can be frequently comprehended, without great dispute, although it is not reducible to any materially visible pattern. The same material configuration can be used in a different circumstance to imply an opposite meaning. The materiality of a sign does not control or determine what it signifies. Signs are much more than matter.

The thought process conducting the inference of the meaning is beyond our senses; and that takes place in many examples of daily life. Consider, for instance, the goodness of a friend is totally outside of our immediate perceptual apparatus, and yet it does exist: based on previous interactions with our friend, we expect the same goodness in the future. In *De Fide Rerum Invisibilium* (1, 2), Augustine asks: "With what eyes do you see your friend's will toward you? For, no will can be seen with bodily eyes." How can one be confident that one is in touch with the mind of others, which do not allow any direct glimpse? It would be unacceptable to presume that materially perceived deeds or words from your friend are his goodness. The goodness of a friend – his good will – is believed, although it cannot directly be seen or heard.

The physical universe is saturated with indications that matter is not everything. Fleeting events of the past are incessantly transformed into immaterial memory, stored as images within us, and retained with great power, defeating what is materially transitory. The gift of memory is bestowed across the created universe; it is seen in the behaviour of birds, beasts, and other animals; otherwise, how could they rediscover their dens and nests? Augustine declares that this is not mere habit, but an influence of immaterial memory (*conf.* 10, xvii, 26).

Material patterns, whether made of sight, touch, sound, smell, or taste, mean something when they are united to that which is not reducible to the physicality of the signal (*doc. chr.* 2.1.1). In cosmological terms, and considering creation as the action of God's will, the material world is always in relative confluence with the divine source that defines what is mind in every living thing. For humans, mind takes the form of knowledge, memory, and will; and because it is widespread in humanity, the intrinsic meaning of delivered signs (*signa data*) can be commonly shared when it comes to the minds that receive them. A sign is not a merely arbitrary discontinuity. It can serve and be understood as representation for the communion of minds because it is not sheer dissimilarity.

Signs are mediators, but where did this mediation come from? Augustine argues that God granted us the mediation of mind and matter not only as human capacity, but also as the concrete and supreme gift of sacrificing His Son, the incarnated Word, to save us. Hence, among all signs, the utmost attention should be given to the narrative of Christ's life in the Scriptures; Christ's life is the supreme sign, the absolute model of human beings; for the Son is the only sufficient mediator between God's perfection and imperfect humanity (*civ.* 9, 15). The ominous presence of mediation is also evident if one considers the human self as a continuous interaction, relating being, knowing, and will: "For I am and I know and I will. Knowing and willing I am. I know that I am and I will. I will to be and to know" (*conf.* 13, xi, 12). A sign is a material existence in the order of being; it both conveys knowledge and expresses will.

The confluence of being, knowledge, and will is another one of God's bequeathed grace to humanity; and the split of these three aspects of the self that are essentially continuous indicate God's action. It may well be that the divine dominion is unreachable and unseen from the our limited perspective as finite beings; but something of the divine still flickers inside us, not as direct and stable light, but as a sign of our lost similarity with the divine order (*similes deo*). Beyond Time, beyond the precarious and the unstable, having crossed the narrow paths of the region of dissimilarity, life is fully achieved. Signs flourish in the material world and are blended into the higher unity of the Divine Verb where Being, the Good, the Truth, the Order, and God repose. That ought to be the aim of all who aspire to a perfect life after death.

The task of anyone's life thus requires the transcendence of the precarious natural world; we should always be in search of what gives order to human impermanence; for we are happy when we find the order that leads to God (*ord.* 9, 27). Although it is not all evident and ever clear, the presence of order is everywhere, even in apparently contradictory signs; while ineffable, order is hidden in the smallest details of the Creation that always point to the Cause of all causes. No joy or happiness could be greater than this kind of understanding. If it was exhilarating for Augustine to realize, in his days at the Cassiciacum retreat, that the discontinuous sound of water flowing was really caused by leaves clogging the channels of the house, would it not be wonderful to grasp – albeit for a short while – the ordered secret of the Divine Providence?

3.7. Which meaning?

Everything that occurs in the material world is therefore a reference to the might and power of God Who responds to all the created. Acknowledging the beauty, the richness, and the appropriateness of this world, one is a witness of the Divine mercy. The reality of the world – brimming with signs – is an indication of God's existence (Gilson 1943). Devoid of divine foreknowledge and astray in a universe where the multiplicity of possibilities would dazzle any human mind, human beings are powerless and ignorant. So, life must never be pursued just in accordance to the conventions of humanity. Life ought to follow the standards of God. To choose man instead of God is a disastrous preference because it represents the inevitable option for falsehood. If not measured against the standard of divine truth, humanity, sin, and falsehood merge into a despairing union. For Augustine, all aspects of human existence are, as one would expect, an initial affair of choice and will. God created the world as a free act of His own will, and – at the same time – granted human beings the right to exercise their will fully although the grace brought with it an obligation: to act in obedience to God's law.

Analyzing the narrative of Genesis, when the Fall of Man is described, Augustine contended that God's command to Adam and Eve not to taste the forbidden fruit was just a test of their capacity to resist temptation and obey their Creator: it was a demand of obedience for obedience's sake. God's commandment was the most forceful way of teaching that disobedience is a great evil (*Gn. litt.* 8, 13), the cause of all iniquities. Failing with comply to the Divine demand, and being incapable of surrendering their will to the Supreme Being, Adam and Eve committed the greatest of all faults, so grave that it has haunted humanity ever since the early days of Creation; and it will be forever with us until the crossing of the threshold of death when bliss or damnation awaits us all. In one of his earliest books, *De Beata Vita* (9, 19), Augustine admonished his brother Navigius, reminding him that the one who seeks God, obeys His will. In Augustine, everything begins and ends in the will.

It is impossible to conciliate the obedience to God's will with the pursuit of a purely secular life. Such alternatives are as contradictory to one another as the flesh is to the spirit. The will of a Christian is never completely free. In the Christian community, it is unacceptable that "some live by man's standards, other by God's (*civ.* 14, 4.) Sets of standard values define the character of the cities in which human beings inhabit. Just as the erratic meaning of a sign is fixed in relation to abstract sets of values and beliefs, each city has its own rules and contracts. To live exclusively by human standards means to mingle among pagans and churn in sacrilege; the best option is to dwell in the heavenly city of God.

The choice exists, but how to exercise it? For Augustine, the reply was, even before his conversion, immediate and obvious, although it posed subsidiary questions difficult to disentangle. Humans have nothing but signs from which to start their quest. God delivers incessant signs (*signa data*) because man has fallen; and now we must restore what was destroyed by the exercise of Adam's perverse will (*mala voluntas*). For that reason, we are adrift in a world permeated with what appears to be conflicting signs because we have lost contact with the source of intelligence and salvation at the moment when Man's will clashed with God's instructions. The outcome of such disobedience was the condemnation to a life saturated with gloom, ignorance, darkness, death, pain, suffering, and deprivation. The price has been enormous; otherwise, humanity would be filled with happiness, living free from death, deception, and distress (*civ.* 14, 25). The impact of the Fall is incalculable. The Fall is the cause of all imperfections of human life. Because humanity has fallen from the grace of God, ignorance is everywhere: more often than not, signs are opaque to what they represent. The spectacle of Creation seems to be articulating something difficult or impossible to comprehend. The overcoming of the hopelessly disjunctive nature of signs demands incessant interpretation, indeed resilient dedication, frequently beyond the capacity of most individuals.

Human disobedience is the source of all ills and sins: before even committing an evil act, human transgression was preceded by the possibility and then the actuality of an evil intention (*civ.* 14, 13). Intention is the foundation of human existence. Because of Adam's ruinous choice, we are immersed in a legacy that has been haunting us since the Fall in the Garden of Eden: alone, using our limited means, we do not know what to do to save ourselves.

Drenched in ignorance and guided solely by human capabilities, how can the human mind identify, yearn for, praise, and subsequently revere the acts of the Divine Providence or even admits to their existence? The only choice for the human mind is to attempt to recognize analogies between dissimilar events. And that in itself – without a guiding light – can be a feeble option: the discovery of analogies always demands more than the mere identification of similarities. The human mind should hold evident, absolute, and guiding standards that would not just produce similarities, but would also assess if one similarity were false, while the other was correct. Otherwise, everything would blur into nothingness, where something is similar to everything else, and no appraisal of actions is ever possible. When producing analogies, and simultaneously holding a standard of judgement, the human mind should generate two types of analogies: it can put forward positive analogies, thus recognizing the common points among objects and facts as well negative analogies that reveal their differences. At this point, analogy is both productive and judicative. Comparing positive analogies, the mind comes easily to the notion that if regularity is observed, it is because there is a force at work, a veiled order binding separate events. The mind then leaps to the assumption that the same explanation is applicable to all cases. This is what is called a rule. Rules provide the way to identify the proper meaning of signs. The idea of orderliness comes from this basic assumption, present in all kinds of knowledge, whether strictly material or indicative of an ordered Divine Providence. What kind of rules are the ones Augustine considered important?

Coherently, the idea of rule in Augustine is drastically theological and differs considerably from the trite admission that human societies are to a greater or lesser degree organized according to normative principles. Since his students days in Carthage, Augustine was aware that different regions of the world and other historical moments had varying norms that were adequate to their place and time (*conf.* 3, vii, 13); but it was clear to him that the demand for Justice and Truth ought to be found beyond the circumstantial and transitory moral codes of each society. Therefore, although agreeing with Cicero's statement in *On the Republic* (*civ.* 19, 21) that the sense of commonwealth in a society is an effect of the just nature of its rules, Augustine counterargues that there was never a real commonwealth in Roman society. Just and appropriate rules are the condition for a just society, but how is justice properly implemented if the creators, the interpreters, and the executors of rules are hopelessly fallible? Human justice is inevitably made in the dark. The judges have no access to the conscience of the ones who are on trial (*civ.* 19. 6); therefore, human judges have assigned themselves an impossible task. No evidence gathered is pure enough to ensure that the defendant was justly condemned or absolved. Torture a witness to produce evidence and still one does not know if the extracted proof or for that matter the confessions were not made to avoid the overwhelming pain of physical suffering. Augustine observes that the ignorance of the judge is a calamity for the innocent. At odds with an idea commonly held during his lifetime, Augustine presented a compelling argument against the use of torture to obtain proof that would condemn the accused of crimes: not just their implementation, but also everything about human rules could be problematic. Even if created according the best possibilities of human rationality, the outcome of rules had to be unsatisfactory: human reason has its limits (*c. acad* 3, 20, 43). More important, rationality could not betray the authority of Christ, the true and exclusive mediator of the dissimilarities and disjunctions between the human and the Divine Order.

The starting point of Augustine's moral and theological semiotics is to be found in the Christian narrative of the Creation. While the human mind is analogical because of its origin in the image and likeness of God, without admitting its limitations, without reference to God's will, humanity will be lost. All that exists in the ladder of Creation ought to be organized hierarchically according to God's nearness. Only respecting the hierarchy of Creation it is possible to control the otherwise spinning conclusions of the human mind. With analogy, an individual may grope in the direction of God, but analogies are simply not enough without the correct authority, the adequate and truthful principle to judge them. Humanity was given dominion over fish, birds, cattle, and creeping things, but man has no authority over the lights of heaven (*conf.* 13, xxiii, 34). It is not on signs and their mere indication of possibilities that full meaning resides. Meaning is in the norms that created the signs, for norms have a firmer ground than representations. Without reference to the rules that created them, signs can be deceptive, and hopelessly incomplete, in the same way that individuals cannot find fulfilment, wisdom, and happiness away from their Creator. At this point, it is evident that Augustine does not reduce norms to human conventions. True rules – moral and semiotic principles – are a Divine legacy, never an abstraction from the commerce between human beings. Restricted to their immanent nature, without reference to a transcendent realm, signs are precarious; they breed insecurity and infelicity: one cannot be happy without a sense of security (*civ.* 10, 30). Not surprisingly, before finding the key to becoming a Christian, Augustine's personal voyage toward conversion in *Confessions* was riddled with anxiety, anguish, and unhappiness.

3.8. Signs and things

Later, as a prominent Catholic bishop in North Africa, Augustine would produce an immense body of treatises, letters, and sermons with the intention of serving as indicative signs to others. By then, the rhetorical purpose of producing pleasant and agreeable speeches had been finally subverted. All signs, and most specifically language, were not narrow means of representing reality, but "a system of pointers that might lead to inner understanding in the reader or listener" (McCormack 1998: 57). Without perceiving the clue that would make him a definitely converted Catholic, Augustine's inner mind scurried along gusts of signs, each pointing to reversed directions. There ought to be a track out of this whirlwind, he must have thought, but no help came to him.

Indeed, help would never come from the immanent and strict domain of signs. To function as pointers, signs have to carry more than their materiality; and that is why Augustine segregated signs that are disseminated all across nature, thus overflowing our senses, which he called *signa naturalia*, from another kind of signs, the *signa data*. In *De Doctrina Christiana*, Augustine traced the differences between such signs: *signa naturalia* are signs that merely exist materially, formed from restricted relations of cause and effect, and thus quite akin to what Peircean semiotics would call indices.[7] *Signa data* are the ones that have the means of acting upon the senses to allow the advent of an idea to someone's mind. If the understanding of sign processes remained tied to the strict sensuality of natural and indexical signs, an important and central dimension of Augustinian semiotics would be missed, namely the conceptual and the mentalist characteristics of the sign patterns exchanged among humans. For Augustine, whatever comes from the external world is inferior to what is internal to the human mind (*ab exterioribus ad inferiora, ab interioribus ad superiora*).

Natural signs express nothing but material presence, whereas *signa data* indicate what is not rigidly present in the materiality of the representation, thus pointing to dimensions other than matter: they manifest an absence. To indicate absences, signs have to bear a mixed composition. Layers of elements of distinct and frequently incongruent or disjunctive nature are placed over each other; and that is why human languages are simultaneously concept and acoustic matter as well as vehicles of illusion, fiction, or deception. The gap between semiotic expressions and their objects is not in itself deception. It is rather a condition for lying; deception and lying come from the contradiction between the sign and the intention of the transmitter. The liar has one idea in mind that does not correspond to what is expressed in words or other signs. A lie is more than a statement that does not correspond to reality. Fiction and hypothesis are not lies. What demarcates a lie is the intention of the liar: "a lie is false statement with the intention of deceiving" (*mend.* 4, 5).

Here, when thinking of human languages, one must be careful not to infer the unwarranted conventionalist equivalence of *signa data* and conventions, as it was done in Robertson's translation of *De Doctrina Christiana. Signa data* are not at all conventional signs. In *De Musica* (6, 9, 24), Augustine recognized that names are imposed by conventions; but the conventional trait of naming does not fully explain the complex relationships between signs and what they signal. Augustine declared unequivocally that names are not important; what matters is their meaning. The inevitable inference is that meaning is not to be reduced to conventions. Meaning is not an actual property of representations, but the indication of the internal dimension of whoever delivers the sign.

Intention is at the origin of language. Nothing can precede the will and the intent to communicate something to someone else. In the initial pages of the *Confessions*, soon after the narrator addressed the Creator asking: "Who are you, my God?" (*conf.* 1, iv, 4), Augustine stated that no reply would be found if he did not truthfully intend to find an answer, if the question were a vain and vacuous display of loquacity. For the question to be meaningful, the narrator should truly intend to speak to God. The possibility of asking this question or making any other inquiry springs from God's grace as He bestowed an inward gift upon the human mind. Consciousness and intent can thus flower without the need of mastering conventions. As soon as consciousness arises, the child begins to impose his or her will on adults through the display of sounds and gestures that indicate the infant's selfish wishes. The cries look similar to anger; mumbling look like dissatisfaction; but they do not have an analogical link to the mental states of anger and dissatisfaction. The signs simply point inward toward the child. Much before becoming acquainted with linguistic conventions, through the delivery of such preconventional signs, the child tries to impose his or her will on adults. The form and the goal of the signs are no more than the expression of the child's will. It is simply too early for any child to understand that adults communicate through conventional signalling. The infant's signs are forceful not because of any form of collective representation; their force comes from the child's will. Although cries can also indicate pain and discomfort, enraged tantrums – a mode of bullying – are demonstrations of willpower. If the adults concede to the child's whim, the infant's will has prevailed. If not, weeping is another pronouncement of the will: it is a willing act of revenge (*conf.* 1, v, 8). The child displays will and intention, even if he or she is not yet capable of mastering a set of conventions

Meaning is what moves the users of signs, and thus it is akin to the free-will that guides human destiny. Although useful as shortcuts, conventions are arbitrary and typical of human limitations. From the human perspective, conventions are inevitably changeable and shaky; but another dimension of rules is revealed when one admits that while human conventions do change, the Word of God abides unchangeable. It is untenable to think of God's Words in terms of human expression. Human words are fleeting, movable, and temporary. Time and evanescence are impoverished human characteristics, quite inferior to God's omnipotent capabilities, for only He knows how to be active while at rest and at rest in his activity (*civ.* 17, 18). The differences are startling. Fulfilling his role of Creator, God begat the Word (*f. et symb.* 3, 3). The Divine Word therefore encompasses the human word: human beings can do no more than form representations; humans are not begetters.

If the Augustinian system of ideas consistently evaluates habits (*consuetudo*) as negative factors in human life, for they are the source of obstacles to surmount when the individual searches for God, why would conventions be any different? Conventions can be even more pernicious; they form habits, granting them social legitimacy; even though they should not be necessarily revered. At best, conventions supply the conditions for sign making and are basically the starting points of human communication. The totality of meaning has to be something altogether different from conventions; and yet, if conventions form linguistic signs, it is because the formed being of linguistic expression is determined by antecedent rules of construction.

Intentional meaning precedes conventions. To mean something is indeed to display unequivocally what the transmitter intended; and to identify a meaning is to recognize the delivered intention. In its moving of signs and users, meaning is definitely mental intention. According to Augustine's speculative argument, angels, for instance, do not communicate with tangible corporeity; their communication is attained through a kind of total vision, addressed "not to the eyes of the eyes of utter speech, not to the ear from without, but within the soul of man" (*ench.* 15, 59). For Augustine, angels are divine messengers, and carriers of the providential will of God; they convey with their messages the full will of the Creator. Will is a grace that the Creator bestowed upon them and upon human beings. Will defines all aspects of God's Creation.

As meaning is reduced to will, the source that granted our will – God's grace – should not be excluded from the acts of signifying. Conventions are not the determining level in the generation of signs. For that reason, Augustine declares, in *De Doctrina Christiana* (2, 2, 3), that animals also have signs that indicate their appetites; but appetite is a negligible mode of intention. The expression of bodily will and emotion is what the rooster and the dove do with their cries, a process similar to the human expression of pain. All such *signa naturalia* are material occurrences without complex intention to mean something, while *signa data* signify because they transcend their corporeal nature.

Whereas *signa naturalia* merely specify the exterior of bodies, *signa data* point toward the interiority of the sender. *Signa naturalia* are not given or intended: they happen; they do not point to the innermost chambers of the one who has emitted the sign. Although *signa data* are essential to the establishment of a community, to identify them with the collective facet of conventions is mistaken. Giving precedence to conventions, one would be expurgating the component of individuality that is so central to Augustine's moral semiotics. To give a sign is to convey a particular, singular, and individual intention; it fulfils a desire and embodies a conscious wish to signify; it is the show of one's heart; and that is why *signa data* cannot be strictly collective and conventional. The idea that the basic condition to deliver representations is that they should be predominantly collective and conventional would be a much later and questionable addition to the understanding of signs (Engels 1962: 373).

Conventions do give rise to words, although words cannot be identified totally with signs. Words are a type of sign. Signs are foremost indicators; and as such, when blended with the variability of the human will, they can point to a variety of objects, ranging from the most pedestrian use of language among humans to the acquisition of knowledge; or even to the portrayal of miracles indicating the presence of God in this world. If that is so, signs refer primordially to signs, leading to higher levels of abstraction. Signs are easily detached from the material world. While erroneous to resume that signs merely take the place of things, signs are means of logical implication; for all practical purposes, signs are inferential tools, and that is why signs indicate universals that are irreducible to any collection of particulars.

With this in mind, Augustine analysed, in *De Magistro*, a sentence of Virgil in the *Aeneid*: "si nihil ex tanta superis placet urbe relinqui" (if it please the gods that nothing remain of so great a city). He wondered how could "si" (if) signify what does not exist, and even more puzzling, how can the word "nihil" (nothing) ever be used as an actual reference? "Si" (if) is doubtful conditionality, and the word nothing ("nihil") cannot be the reference of anything; otherwise, it becomes something, in blatant contradiction with the attribute of nothingness. No collection of existing objects can provide the ground for inferring "nihil" (nothing). It seemed obvious in the dialogue *De Magistro* (2, 3) that doubt and nothingness pre-exist in the mind.

"Si" and "nihil" are fully meaningful because the mind relates a given sign – the linguistic term "nihil" – to a mental representation that is not necessarily the indication of an actual thing; and then, through equivalence, the word has meaning. An offshoot of mental and logical equivalence, synonymy – in fact, a case of semiosis – is both a product of the mind and a self-referred representation. The ancient grammarian and the professor of speech in Augustine would contend that synonymy is a central feature in defining the meaning of a sign; but can the meaning of words be represented with other words alone? Apparently so, although that is not the case: names can be pointers, as when one indicates with a gesture this specific wall (*paries*) referred to in a conversation (*mag.* 3. 5). A gesture can clarify an enigmatic representation; and yet is the pointing of a finger toward an object not a gesture, an occurrence that is obviously a sign? As long as it can be materially perceived, any physical entity (*res*) can be turned into a sign; although the physicality of signs is insufficient to indicate the transcendental.

Transcendence is manifested when the mind moves away from what it is present to the senses and progresses into higher and higher abstractions, through the establishment of equivalences not just of signs and things, but also of signs and signs. Pointing a finger forges the relationship between a representation and a thing, thus reducing the disjunction between *signum* and *res*: it is a mental construction. Therefore, if pointing is a sign that clarifies a representation, this is the case of signs referring to signs, signifying mutually one another. If Augustine's argument were arrested here, the conclusion would be that nothing can be shown without a sign, but when I walk to indicate what the word "walking" means, I must do it so that it is explicitly an illustration. One can produce meaning by telling or by showing, but that would pose another challenge: how to display a sign that would indicate walking in haste and not be confused with slow walking. One can underscore haste moving excitedly the arms in an almost caricature. A clue must be given to the mind receiving the sign. As always, the meaning would not be in the sign, but in the mind interpreting the display.

3.9. Knowledge and semiosis

However, if signs have the primary function of conveying knowledge, the ceaseless reference of signs to signs presents a major dilemma, as is evident in the case of infinite semiosis. Both types of semiosis – synonymy and definition – are irrelevant, indeed positively detrimental if signs are to be deemed cognitive resources. What occurs when one looks up a definition in a dictionary, or when one provides an equivalent representation to pin down the meaning of a sign is the plain triggering of a nominal regression, an effect of signs referring mutually to other signs. In this whirlpool of signs endlessly referring to signs, knowledge is reduced to a linguistic game, an empty verbiage, without a standard of judgement and neglectful of any commitment to finding the truth. Truth is not strictly semiotic; it is neither in a representation nor in their objects. As an assessment that comes from the relationship between signs and things, truth is an extra-semiotic attribute.

Augustine could never adopt with ease this conception of truth and meaning as purely semiosis: it was an exceedingly secular response to a harrowing theological question.

Nonetheless, in the dialogue *De Magistro,* while Augustine taunted his son, Adeodatus , to accept the reduction of meaning to semiosis, the young boy observed that a loathed word such as *coenum* (filth) was repugnant not because of the properties of the sign, but because of its referent. The name *coenum* – the young Adeodatus points out – is quite close to the desired and revered *coelum* (heaven); the two names differ because of the sheer change of one letter: "n" for "c." And what generated loathing was not the sign but the thing signified. Adeodatus' remarks then led to a discussion about what is superior: the signs or the things. To produce a proper response, Augustine and Adeodatus found it necessary to define the conditions of superiority. Augustine claimed and Adedodatus agreed that the idea of superiority springs from the corollary that whatever needs another to exist is inferior. Therefore, if one hears a sign with greater complaisance than one perceives the thing (*mag.* 9. 25), and one uses the sign not to indicate actual filth, the conclusion can only be that the sign is independent from the thing, as well as superior to their objects.

What is the purpose of using a sign? Signs are used to signify; for without meaning, teaching and knowledge would be impossible: "the signs exist because of the cognition, and not the cognition because of the sign" (*mag.* 9, 26). Examining the implications of this argument, father and son jointly work out that teaching has to be superior to signifying, or more precisely teaching is superior to naming.

In *De Magistro*, the constant assumption is that the exchange of signs should have a purpose, and that the higher purpose of the delivery of signs is to teach so that someone can learn. What, then, is it that one teaches and another learns? One should only teach if one knows. Knowledge is thus superior to names and things.

The issues now become: How does one know? What is true knowledge? The initially rigid distinction between signs (*signa*) and entities (*rei*) cannot easily hold. Are signs really superior to things, especially after admitting that signs address things that must be known? In this circumstance, signs become dependent on things: "no sign is known if it is not known of what thing it is a sign" (*trin.* 10, 1, 2). Puzzling as it may be, and in reference to the established criterion of superiority, signs have to be inferior to their objects. How can signs be inferior to things, when they seemed superior not awhile before in Augustine's dialogue? The path of arguments in *De Magistro* does not seem to lead to a satisfying solution. The dialogue went backward and forward without presenting a definite reply to such a basic question. What were Augustine and Adeodatus doing? Were they trying to demonstrate that teaching and knowing are essentially riddled with paradoxes? Was their goal to show that dialogues are impossible? Why take the trouble of using a dialogue to show that instructional dialogues are useless?

The Augustinian trail of thought in *De Magistro* appears to illustrate the progress toward a hopeless intellectual dead end; but it may well be that the sections of the dialogue on signs and things were no more than "a tactical paradox" (Madec 1975: 71). The whole point could be not to determine the viability of teaching, but the conditions of its possibility. Thus, by extension, what the dialogue really intended was to demonstrate the impossibility of a completely secular knowledge. In the same way that one should not reduce the delivery of signs to their external materiality, truly meaningful communication is not that which occurs horizontally during the commerce of human beings, exchanging signs with one another. In the case of a strictly secular viewpoint, signs, things, and knowledge move in circles, leading nowhere, as if pursuing their own tails.

Strictly secular knowledge cannot solve the impasse between signs and things. Another element must be added to break the impasse. Augustine would argue that words and signs are simply not enough to indicate true and relevant knowledge: to hear or to receive signs is rather useless if one does not know what the word or sign represents or signifies. The knowledge of things completely determines the knowledge of signs; so when signs are delivered, the receiver must already somewhat know or accept their meaning. That is not exactly learning. It is rather an evocation or more accurately a remembrance (*mag.* 11, 36). Augustine's *De Magistro* stitches an ingenious paradox, whose purpose is not immediately clear to the reader.

For Augustine, that paradox could only be solved through admitting the special status that believing has over understanding. If meaningful communication is not horizontal, it is definitely vertical, involving the inward dimension of human beings and their supreme Creator. The discrepant order of the human and the divine is breached when the individual surrenders his or her will, believing and accepting Christ, as the Teacher from within, the sole mediator both son of human beings and of God, the supreme gift of the Creator to humanity. That is how Augustine described *De Magistro* in *Retractiones* (1.11): it was a book not on the discussion of signs, but an exploration on the true nature of teaching. It is therefore legitimate to state that the book's whole weaving of concepts about the relationship among signs, things, and knowledge is purely accessory to a more central thesis: that the only true knowledge comes from God, as it is verbally expressed in Matthew (23: 10): "one is your Master, the Christ." It is also worthy of note that although not the last work under the dialogue form to receive Augustine's final touches and revisions, *De Magistro* was the last dialogue that he started, granting to it an aura of culmination. It is also quite telling that in *Retractiones* Augustine makes no amends or adjustments to any part of the book he began writing around 389-390, after his leaving Ostia and returning to North Africa. This was a time when Augustine was probably also writing the *Confessions* and being completely immersed in the overwhelming duties of a Catholic Bishop.

Augustine's claim about the precedence of beliefs was not merely theological and religious. He contended that his position was predominantly epistemological for believing is superior to understanding. It was obvious to him that the abandonment of all modes of scepticism ought to be complete. Putting doubt at the origin of any intellectual quest is a major hindrance to cognition. Unshakable and absolute acts of believing should be at the "threshold of faith" (*ench.* 7, 20). Compare believing with understanding and see that one believes everything that is understood, but one does not understand everything believed. Believing also fully satisfies the conditions of teaching and meaning in *De Magistro*. Believing is spiritual and internal – not external – thus a product of the mind. Augustine was quite convinced that it was mandatory to shake up any submission to what is external. The voyage of the believer must be always inward. No actions –

even the best deeds – can be judged in themselves. The meaning of an action comes from submission to God's commands.

God is not external to humanity: He is the ultimate internal dimension of human beings. The will and power of human beings is dependent upon the will of God (*civ.* 5, 9). Human incompleteness springs from the Divine totality. One can only act virtuously if in accordance to God's commands (Dihle 1982: 131). Imperfect human liberty craves Divine omnipotence. Thus, since human beings cannot have foreknowledge of their destiny, being prisoners of a limited and distorted perception of the future and devoid of a full grasp of the consequences of the exercise of their will, it is only prudent and correct to submit humanity's feeble and contradictory liberty to God's norm.

Submission must be total and fearless. Fear is an unacceptable sentiment in reference to God who has demonstrated his love for humankind sacrificing His son to redeem us. The cross is the supreme gift of God to humans. Therefore, it is upon humanity to acknowledge such a gift with the same loving intensity, and, contrary to Adam, reciprocate God's bequest with acts of total obedience. As it is always the case with the deepest mysteries of life, it is a matter of believing, and believing – indeed taking the leap of faith – is surely a matter of the will.

With believing, God's supreme grace is kept at the core of Augustine's system of ideas. For a Christian, believing requires conversion, adoption of the faith, as well as the surrender of one's will; for Christianity is quite distinct from other religions bonded to ethnic and social ties. It must be remembered that in Augustine's days Christians were baptized because of their conscious choice, not because of the biological origin of the individual. Although a devout Catholic, Monnica could only look forward to Augustine's conversion; as a conscious, personal, and individual decision, her son would be baptized at the late age of 33 years old. Late baptism was a well-established practice among early Christians.

For a religion strictly based on private choice, baptism at the age of reason also reiterated the radical division of nature and culture reflected across Augustine's works and presented in his basic typology that segregated natural and given signs. To a Christian, the separation of nature and culture must be fully achieved. Augustine felt that, because of all this, sex ought to be excised from his life; and reduced to the intention of procreation for the rest of the brethren. Furthermore, Augustine found the solution to the paradoxical quandary of *De Magistro* in the words of the Biblical prophet Isaiah (*ep.* 120). If you do not believe – said the prophet – you will not understand, at which point Augustine was arguing for the centrality of the

Scriptures in the formation and the shaping of the conduct of faithful Christians.

3.10. How and where to find the norms

For Augustine, faith precedes rational understanding. In *De Trinitate* (15, 2, 2), the formulation is unwaveringly lucid: "faith seeks, understanding finds" (*fides quaerit; intellectus invenit*). More than the interior light of truth, faith requires the acceptance of the authority of the Church guarding the word of God. Christians must live under the instruction of the supreme inner teacher, the shaper of lives, the guide to human action. It is therefore demanded of the faithful that they ought to place themselves in hands of Incarnated Word, abiding by Christ, the mediator of the disjointed realms of the human and the Divine. Although that may be an inevitable suggestion, a problem remains: How can one find the Word? First, one should turn inward, searching for the unseen Creator, which is never fully expressed in the beauty and majesty of nature. Essential attributes of the Creator – perfection and permanence – are not in nature: nature is corruption and change. The way to reach the Divine is not through fallible human knowledge but through the faith that truly intends to know the Divine. True faith is unchanging and provides the steady anchor of human understanding. Understanding can be within the reach of the very few, whereas faith is accessible to all, limited or not in their intelligence; and not only that: human intelligence devoid of faith only weaves doubts. To cull doubts – the source of anxiety and suffering – one ought to bow down to the authority of the Church that determines the contents of the faith. That may be against the force of human pride, but humility is necessary to salvation. To save the soul, the individual has three alternatives: "first humility, second humility, third humility" (*ep.* 118).

Lack of humility generates disobedience, as it originally happened in the Garden of Eden, when the serpent tempted Adam to act with the pride that is a trait of god-like entities. The narrative of Genesis, both in the literal and the figurative sense, shows God teaching to Adam that neither he nor his offspring will ever be gods. At this point, humanity fell from grace and was doomed to dwell in the opposite realm of Divinity. That is a lesson to be learned throughout all humanity's life on Earth. It is a teaching that reaches goes beyond Adam. Adam may have personally learned the consequences of his act, but the story of his arrogant disobedience requires that humanity stop committing acts of pride (*Gn. litt.* 11, 39, 53). The Biblical story is a normative illustration that ought to inspire all human beings. The acceptance of the Scriptures and the creed in the perfect sanctity of God's Word supplements the mere moral adoption of the virtue of humility. What Augustine professes is that the choice for a humble life is not just a moral and logical decision on the part of Christians. It is an act of faith far superior to any rational understanding.

As faith is possible without reason, so reason can function without faith. Authority (*auctoritas*) and reason (*ratio*) are isolated means of acquiring truthful knowledge although their hierarchical relationship is obvious considering how belief is deemed superior to understanding. The authority of the faith should command rational decisions. Therefore, if there is no understanding without belief, and belief can be perfectly held without understanding, this is the conclusion; I believe so that I can understand (*credo ut intelligam*). That is a tenet consistently active in Augustine's writings. In one of his earliest dialogues, written just after his definite option for a Christian life, in *De Ordine* (9, 26), he pronounced that authority comes first. During his long career as public thinker, Augustine's intellectual efforts amounted to attempts at grasping what is the intelligible and ontological, albeit reduced to the Divinity materialized in the Biblical narrative of Christ's life among humans (Holte 1962: 361). To do that, Augustine indicates that human life is caught in fields of opposing forces. It is *ratio* against *auctoritas*; signs detached from things; what signifies contrasted to what is signified; the external as opposed to the internal; and it is only in relation to a Divine Norm that these contradictions can be appeased.

Although under a theological and strictly non-secular guise, the main tenets of conventionalism are present in Augustine. The idea of human communities living under shared rules and principles is at the core of Augustine's social theory; but that type of peace and union is transitory and unsatisfactory; and it is no wonder that human societies are torn apart from time to time. The unity of humanity is only harmonious in peaceful obedience to the Scriptures (*civ.* 14, 1). For Augustine, the Divine norm is a convention, but of the strongest form.

Humanity has an unbreakable contract with God, to whom everything is owned. Not just our life was given to us; but also the Creator gave us His son, the redeemer who released humanity from Adam's bondage: like no other religion, Christianity postulates a doctrine of eternal forgiveness that erases the most ancient past. The soiled origins of humanity can be washed away; but only through Divine help. The human debt to God is immense.

Throughout Augustine's massive body of work, conventionalism takes a surprising and radical turn. It is not just an element of social differentiation. It is altogether unlike the Homeric conception of heroic and tragic individuality that haunted Antiquity. For Christian communities, diversity is deemed irrelevant if compared to the peaceful bliss of a homogeneous life within the heavenly city. The rules of the city of God are radically different from those of the city of men. Strife and competition are the hallmarks of diabolical pagan gatherings whereas the city of God is united in its love for the prescriptions of the Scriptures. While salvation can be universal, the means to attain redemption are available to every single human being who surrenders to the instructions of the Scriptures. Conventionalism and collectivism dissolve into one another under the severe seal of the Divinity. In the city of God, existence is an act of capitulation. The higher will of being with God refrains the lower will of the libido.

Augustine did more than uphold the main tenets of conventionalism. He intensified and expanded them to a point never seen before in the Western canon. In the Christian social conception, conventionality still separated humanity from the natural world. What is particularly Christian is the notion that conventions ought to be found in the Biblical narratives. The Scriptures provide the stern guidelines for coordinating the behavior of selfish individuals. To overcome the destructive and prideful traits of humanity, submission to the sacred Biblical norms is compulsory. For both pagans and Christians, social norms may appear to take an arbitrary fashion, because they are voluntary determinations; but what makes the city of God superior to the pagan world is that the will presiding the norms in the heavenly city expresses God's will.

The social effectiveness of conventions always depends upon reciprocity; but restricted human reciprocity – being horizontal and between individuals of the same kind – inevitably yields wickedness and violent power struggles. In the city of justice, however, reciprocity is vertical, and therefore capable of creating lasting harmony and stability. God, the enforcer, is present inside every dweller of the heavenly city. The presence of the most powerful enforcer is precisely what had been missing in all secular social theories until the monumental *De Civitate Dei*. After Augustine, conventionalism is as potent as it ever could be.

The conventions of the heavenly city are irreducible to written laws prescribing righteousness and appropriate conduct. The prescriptions and the instructions springing from Biblical interpretation request much more than chastity and rectitude. Otherwise, Biblical texts would be mere blotches of ink on pages; they would be what the Pauline epistle to the Corinthians (2. Cor 3: 6) called "the letter." That kind of letter kills; for human laws and inevitably mortal and precarious: they do not follow the eternal spirit that gives life (*spir. et litt.* 4.6).

In Augustinian social theory, no human law makes sense without Divine grace; and grace comprises the spirit of the living God of the Scriptures. Here, Augustine blends inseparably believing and understanding. Christians must know and abide by what is told in the Scriptures because of their acceptance and belief in the supremacy of the Creator. Religious belief entails knowledge, and knowing is achieved by means of believing. Purely secular culture – even the one that Cicero argued for in his attack against the deficiencies of classical rhetoric schools – is simply not enough for living a successful Christian existence.

Life among humans requires definite and absolute norms that curb the inevitable selfish drives of sexual beings. Cooperation and submission are preferable to any mode of individual competition. Discipline and doctrine are mandatory if one aims at a full life in God. The disciplinary benediction of the Spirit embodied in the Scriptures acts over humanity subordinating and ruling the otherwise chaotic life of secular individuals. In that sense, the Christian doctrine is a means of coordination and cultivation. It is culture in the contemporary anthropological perspective of the term with doctrinaire and with a definite historical slant. Augustinian theory of history does not conceive the flow of time as the eternal repetition of the same. To those natural philosophers who contended that "the world disappear and reappear, showing the same features, which appear as new" (*civ.* 13, 14), Augustine responded that the world was like a slope ascending toward perfection; if not, how could humanity be rescued and redeemed from circuitous misery? History is a tortuous path to bliss, beginning as fall, then guided by prophecies that lead to the coming of the Messiah and terminating in the resurrection and salvation of the individuals who have embraced the instructions of the Biblical doctrine. History moves toward a goal because its laws are not concealed but given as rules to guide humanity.

The Christian doctrine and its culture are means to improve humankind. Infused with the spirit of Divine grace, Christian rules are positive additions that will modify for good an otherwise savage state (Kavane 1967: 124). If ordered by shared Biblical conventions, social life can be more than just perdition and pain. History is our common duty, as it is our mission to obey God's universal contract with us. All further descriptions of human societies organized around rules that guide and determine human actions simply revive – under different forms and degrees – the Augustinian dream of an ideal community.

3.11. The light within the heart

However, to finalize a precise analysis of Augustine's idea of culture and conventions, it is indispensable to ponder not only about the formative role of believing and understanding for a Christian community, but also to see what relation these attributes have with the experience and the process of illumination. Believing and understanding are feasible because the human mind undergoes enlightenment by a source that is superior to it (Stock 1996: 160). Deprived of illumination, humans could never achieve believing and understanding, and without the divine enlightenment, Augustine hierarchical conventionalism would be incomplete.

While believing and understanding generate concepts that can be true or false, illumination points the way to truth. The role of illumination in the quest for the Divine truth is not a minor one; it is rather a condition for reaching certainty and God: if the Divine light did not inundate the soul, human life would be desperately meshed in error, contradiction, and foolishness. The truth is never hidden, Augustine states in *De Utilitati Credendi* (8, 20), it is only hidden in the way that it is sought, and that should be taken from some divine authority. So what does illumination makes to the soul? In his masterful exegesis of Augustinian philosophy, Etienne Gilson (1943: 112) carefully answers this question, pointing out what illumination does not do: it is not an intuition of the contents of God's idea. Human knowledge is opaque to God's thoughts. It is as Paul said: humanity perceives God through a glass, darkly.

Illumination is not a metaphor either; it just means that the Light allows for knowledge, both secular and divine. Illumination does not reveal knowledge; or else God would be active and the human mind passive. If that were true, this assumption would make the effort toward individual salvation useless. It is part of humanity's task to actively search and find God. Indeed illumination does not relieve humanity from the quest for God. It is rather the other way around: because illumination exists, humanity ought to find God with the intellect that has been endowed to us. God's illumination is nothing more than a condition for truth under all forms. It is a universal gift of God to humankind; therefore, illumination never leaves us, even when we turn our backs to Him.

Human hunger for truth is a shadow of God's light, even though the search itself may be misguided and lead to nowhere, or to an intellectual dead end. Although illumination is always directed toward truth, it can be temporarily darkened by human beliefs and misunderstandings; but when the individual feel the divine light shining upon him, he or she undergoes a mystical vision.

Mystical visions are surprisingly narrow. They cannot be adequately transmitted with precision to anyone else but the illuminated. The mystical experience invades the individual with the presence of God; but they do not provide concepts that can be shared. Understanding shapes the conceptual structure of any doctrine, while illumination generates the means to judge it: "what your intellect sees in the light of illumination and not by its own light is the truth of its own judgements" (Gilson 1960: 86). That is why it is easily said that the intellect can see whether an idea is right or wrong, good or bad, even before it can assess it rationally. At this point, the mind becomes aware that something entered its thought. The mystic experiences truths that are not merely temporal and transitory, but also comes to the revelation of immortality and eternity (*imm. an.* 4, 6) that otherwise would be foreign to the human soul.

Because illumination precedes and informs understanding, the pursuit and the grasp of truth are under no circumstances solitary. For Augustine, truth comes from within but is only achieved after being revealed in public. The force of truth cannot be held in privacy. That is one of Augustine's most cherished assumptions. If not, why would such a prude man, so conscious of sin and guilt, open up his painful secrets, his bleak sexual yearnings, his harrowing indecisions and self-accusations in many passages of *Confessions* and treatises as well as sermons and letters? To him, gossip is certainly no reason; and an indication of Augustine's discreet tact is that he refrained from exposing the name of his concubine – Adeodatus' mother – and from identifying who was the close friend whose death he grieved so much in the fourth chapter of his literary masterpiece (*conf.* 4, iv, 7–9). While the individual may appear to be alone, even if ostensibly immersed in the most absolute solitude and seemingly detached from his or her fellow human beings that is never the case: the Divine light – which may or not may be seen – is always present.

Illumination is unrelenting. As we are always in the company of God, signs are flashed before us, even if we do not see them. Augustine's *Confessions* lists a myriad of opportunities lost before Augustine's final option for a Christian life. When a young sick boy, "hot with fever and almost at death's door" (*conf.* 1, xi, 17), his distraught mother almost baptized him. *Confessions* describes decisions that ever could never be taken, lusts not quite renounced, signs on the verge of proposing something but never do, or if they do, it is in an infinite manner, incomprehensible to the finite mind and thus incapable of being translated into an undisputed meaning.

When the soul longs for God, its hope is for both a perpetual bond and a perfect communion with the Creator. Communion and companionship are at the origin of everything; and for that reason, because it is the destiny of all, God made immediately a companion to Adam in the Garden of Eden, and yet more important, He made the first man in his image and likeness. No individual is absolutely singular. Even though Adam was a warped mirror, like all mirrors, which need models and objects to reflect, the original man had to exist in the company and the community of others.

Augustine's initial endeavours at illumination failed not just because of his incapacity of refraining from sex; but also and perhaps mainly because the effort was solely introspective. The Plotinian One is less a union than a selfish ascent leading to "an innermost citadel" (*conf.* 7, x, 16), which would easily shelter him from others. Although granted by God, it was precisely *Amor Dei* that was absent in Augustine's trials. Without that, he would not attain successfully the intended ascension from the volatile world of the senses. *Amor Dei* requires a move beyond the sanctuary of the self. Such a love presumes the existence of a divine entity to be reached and somewhat known: Love and knowledge are inseparable: "How can anyone love anything that he does not know?" (Louth 1981: 149).

Only after his conversion could Augustine have an experience approximate to a mystical vision. The process of conversion was an individual choice and yet underlined by the interaction with others. He heard Symplicianus celebrate the baptism of the well-known rhetor Marius Victoriunus; Augustine was then informed of Anthony's of Egypt option for a monastic life; Ponticianus, another North African recently converted to Christianity, told him of members of the imperial civil service of Trier that were suddenly filled with the love of God and, renouncing sex, became Christians. Augustine felt alone and incapable of escaping from his own cage. He could not understand why uneducated people could find God and he did not. He expressed surprise and frustration to his dear friend Alypius.

Then, signs of anxiety vibrated over his body, pointing to the garden of the house he is temporarily living in Milan. He interprets the signs invading him as a physical agony akin to dying. Death is the ultimate solitude. Alypius follows his steps, but Augustine feels no intrusion on his loneliness (*conf.* 8, vii, 14). Augustine's extreme isolation does not allow him to reach for anyone but himself, so he cannot enter into a pact and covenant with God. A contractual alliance (*placitum*) with God stipulates entire submission; and that is impossible to come about when the will is split. This division of the self hinders communion and only aggravates Augustine's sense of remoteness: a divided self is more than ill; it is prone to evil acts. Disorders of the will "are purely internal to itself" (1992c: 49). Cure should come from the outside as a gift from God.

Augustine then enters the garden. His narrative in *Confessions* takes a sudden allegorical form;[8] and he perceives the figure of the dignified and chaste Lady Continence inviting him in. He has found no peace yet; the struggle rages within his heart (*ista controversia in corde meo*). An emotional storm breaks out. He feels like crying. Tears drag him into a deeper well of loneliness. He fears that he cannot meet the colossal demand of reinventing himself. Then, a voice comes from a neighbouring house[9] and exhorts him like a song: "pick up and read; pick up and read" (*tolle lege, tolle lege*). Augustine stops crying as he grants to this sign the solemn meaning of a divine command. He searches for the book that he carried with him. The reason for Augustine's interpretation of a human chant (*cantu*) as a divine exhortation is based on another sign: a similar situation had inspired Antony of Egypt to opt for a monastic life; Antony had read a passage of the Gospels (Matt. 19: 21) and made it into God's encouragement. With the epistle of Paul to the Romans in his hands, Augustine reads the random passage (Rom.13:13-14) in silence; and finds in it a condemnation of unbridled eroticism and the worldly temptations of the flesh. Augustine interprets this sign as God's message to him. For Augustine, what appears to be chance is frequently blindness to grace and predestination. A soothing light bathes him with relief, brightening him up and casting the shadows of doubt away. He feels fully converted to the Christian religion. Indeed it is difficult to call that a sudden conversion. Why was Augustine carrying a book of Paul's epistles if he was not already halfway through the process of opting for a Christian life? The garden scene is less a conversion than a confirmation.

Of course, that could not be all. Augustine goes to Alypius and describes what had just happened to him. Alypius asks to read the Pauline epistle and recognizes in it a veiled message for his conversion, too. Jointly, and much to her joy, they tell Monnica of their decision to be baptized. Augustine also notifies his mother of his vow to pursue celibacy and chastity.

Although it is impossible to probe beyond Augustine's text, the narrative of his conversion illustrates his more abstract theory of meaning. Signs relate to other signs, but meaning is lost if it does not spring from the inside toward a public representation. That movement, however, should also bounce from the external to the direction of the innermost chambers of the heart. The directions – from inward to outward and from outward to inward – are not dependent upon each other. Meaning is not solipsistic; it has a public dimension. The rules prevailing in the heavenly city should be more than just individual norms. They must be public and materialized to anyone who would be willing to search for them in the examples of the Bible. Submission to the scriptural norms reshapes unruly humankind.

Augustine upholds a conception of public rules that go beyond the limits of a specific human society. Biblical rules produce effects over the actual cities of the pagan world; but they are ecumenical, indeed universal. The real formative rules are the ones that do not come from the human kingdoms, but because of their hierarchical position, they can impose order on the secular realm. Without this insight, and incapable of seeing the incidental signs of life in front of him as given by God, Augustine – like so many of us – missed other clues that were displayed to him much before the garden scene in Milan. His mother's dream (*conf.* 3, x, 18; and 3, xi, 19) that he would find his path back to Christianity was just her personal reverie, not a command addressed to him to abandon the Manichean sect.

The most significant elements of the garden scene – the silence, the voice, the Bible, Paul's epistle, the isolation of the individual, the internal discovery made public – were pointers already presented to him in an earlier meeting with Ambrose (*conf.* 6, iii, 3). During one of Augustine's visit to Ambrose, he saw the bishop of Milan reading the Pauline's scriptures with his heart, and never moving his lips. Later, Augustine remarked that Ambrose had preached on the passage of Paul's epistle that referred to the letter and the spirit. Ambrose's reading was not of the letter but of the interiority of the spirit. The event made a vivid impression on Augustine, not only because of the rarity of seeing someone reading in silence,[10] but also because it seemed to him that Augustine had silently discovered "the "spirit" and the "letter as a parallel to the inner" and "'outer' self" (Stock 1996: 54). It was not only the "inner" made "outer"; or the "outer" without authenticity determining the "inner." The unconditional synthesis of the "inner" and the "outer" was for Augustine the decisive goal of the sincere inquiring mind.

Nonetheless, the old bishop Ambrose could not be seen as model for Augustine, despite the admiration of the young professor of rhetoric. Augustine could not see the signs, although they were there visible and evident. Ambrose never seemed to be in a tormented solitude; like Augustine, he had been trained in the liberal arts; he had climbed the social ladder of the Empire and had become a provincial governor; he had abandoned all aspiration of worldly success; he had no *ambitio seculi*; he was the magisterial interpreter of the Scriptures that Augustine wanted to become; he lived in celibacy, and was happy in God. Ambrose, however, was a man just like Augustine. No redeeming example can come from human beings, not even if they have lived an exemplary life. The true example does not come from Ambrose's achievements; it springs from God's words in the Biblical scriptures. The inner faith should meet the outer narrative of Jesus' life on Earth.

After the internal storm (*procella*) of signs in the garden scene that confirmed his decision to become celibate and submit his desires to the Christian principles of renunciation, Augustine publicly gave his name (*nomen dare*) for baptism, whose ceremony was probably performed at the octagonal baptistery built under the Piazza Duomo in Milan. The full circle of meaning was about to be completed. With Alypius and Adeodatus, Augustine was baptized and thus reborn in God; but if this act concerns the rebirth of the baptismal experience, it is necessary to inquire about the destiny of the individual soul that follows death.

Much later, probably in 421, Augustine writes *De Cura pro Mortuis Gerenda* (*On the Care of the Dead*) in response Paulinus of Nola's question about the more advantageous burial site for the dead. Was it better to be buried closer to the shrine of a saint (*ret.* 2, 90)? The tone of Augustine's book is brimmed with emotion, understanding, and compassion for the vicissitude of dying. It does not really matter where one is buried. That was the lesson that his mother's request taught him. The souls of dead inhabit another realm.

In *Confessions* (9, x, 23–25), Augustine records one of his last interactions with his mother. Monnica is not ailing of an identified illness. Her death would come as a surprise to him. They are leaning out of a window overlooking a garden of the house where they are staying in Ostia. They are tired from the trip through the *Via Ostensis* that took them to the left bank of the Tiber (Perler 1969: 135). They begin to speculate about the afterlife and the life of the saints. Augustine does not present a record of what they said to each other, but is obvious that they conversed on the subject.

While the confirmation of Augustine's faith in the garden scene of Milan is essentially solitary, the experience with Monnica is communal. What occurred happened to both of them. Their experience was mutual, although not simultaneous (Starnes 1992). The vision of Ostia is quite different from any Plotinian communion with the One. It is a revelation that comes from the outside to reverberate inside of Augustine and his mother. It is not a denial of the corporeal nature of the world, in any way similar to the extreme renunciation of the bodily drives that gives to sex a pivotal role in the Milanese garden. In Ostia, Augustine and his mother are lifting up their spirit, going beyond the senses, and transcending the limited expression of the discourse, thus eventually coming in silence to the eternal being, which is reached during a moment of total concentration of the heart (*toto ictus cordis*).

The movement of Augustine and Monnica is ascendant; their starting point was the region of dissimilarity and they end up in a blissful state of inexhaustible abundance (*regio ubertatis indeficientis*), where nothing lacks. Augustine and Monnica have reached a region of supreme good; it is the opposite of what he found possible to experience in earlier parts of the *Confessions* (7, x, 16), where the supreme good was a mere intellectual inference, an intention not realized. The vision of Ostia is an insight that cannot be captured through uttered signs, "by our human speech where a sentence has both a beginning and an end" (*conf.* 9, x, 23). This admission confers a degree of existence to a region that is akin to the prior sphere of intentions when the child acquires speech. There is a dominion before and beyond the signs.

Pierre Henry, in this classical study on Augustine's mystical experience, *La Vision d'Ostie* (1938 and 1981), observes that the vision of Ostia is not an ecstasy, a successful attempt at bliss that may have started from his reading of intellectual works. The vision – if that is a sight – is an experience shared with a semi-literate woman; it is an emotional and overwhelming anticipation of the eternal life waiting on Christians. While Augustine used a Platonic vocabulary to record this experience in the *Confessions*, the active presence of Monnica counters the suggestion of an extremely intellectual achievement. It is a moment of reward and retribution for Christians who have lived their time on Earth in obedience to the divine norms of the scriptures.

The celebrated vision of Ostia is many things: it is the object of a description of a moment in Augustine's life; it presents thoughts about the limits of language and signs; it is an admission that the world of senses, the human world, where we dwell, acquires meaning and purpose only in relation to God and His commands. The vision of Ostia also displays Augustine's model of knowing that will persist throughout the rest of his prolonged life as an influential Catholic thinker, and that will be developed and evolved in *De Doctrina Christiana,* his guide and legacy for a Christian life. The experience of Ostia embodies a definite conception of truth. Truth is not to be found in books, Augustine declares in a letter to Gaius (*ep.* 19), written around 390, soon after the vision of Ostia. Truth is not in the author who wrote the words either; it is inside one's mind, illuminated by the light of veracity, an experience removed from any kind of material brightness.

In one of the dialogues of the Cassiciacum, in the *Soliloquia* (1, 4, 9), Augustine compared the senses with a ship that carries someone to his destination. As early as *De Beata Vita* (1.1) – not long after his conversion and quite close to his baptism – philosophical inquiry is described as a port from which one can enter the hinterland of the happy life. Knowledge of God is a trip through stages (*quant. an.* 70-75); but one should leave the vessel behind, and empty it (*civ.* 11, 10). It is only possible to attain the knowledge of God, identifying what is not God (*trin.* 8, 2, 3), and thus in the *Confessions* (10, vi, 8), the path to God is paved with clearly recognized "negatives." God is not the physical beauty of nature; it is not the temporal glory of public life; it is not the same as the brightness of light or the sweetness and odors of flowers and perfumes; it is not food, it is not flesh. This is not the love of God. God is what is left when such things are shed away. The access to God's affirmation is sprinkled with purifying negativities. Augustine's role in the future of Christianity is thus closely tied to his actions as a staunch opponent of what he perceived to be the breaches of faith of heretical Manicheans, Pelagians, Donatists, among others. The polemicist in Augustine is inseparable from the visionary who had the experience of Ostia.

After the vision of Ostia, Monnica died. Augustine buried her there. To be close to her grave was unnecessary: the dead do not participate in the lives of the living. Monnica became memory. She did not guide him by the hand anymore, as she had so frequently done (*cura mort.* 13, 16). On July 28, 388, the day in which the *Consularia Constinoplana* recorded that Theodosius finally defeated and killed the usurper Magnus Maximus (Perler 1969: 197–203), the blockade of Ostia was lifted, and Augustine knew that he could sail back to North Africa. All that he hoped was to be devoted to a monastic life. He wanted to become a writer of philosophical books and to be a *servus dei.*

Like any traveller who departed from Ostia, he boarded a boat that would carry him to the larger ship that would take him home (Perler 1969: 145). It would be the last time Augustine sailed to and from North Africa (Perler 1969: 145): he feared the sea (Brown 1967: 191). As the Bishop of Hippo, he would travel by land, preferably in horseback, musing to himself as the trip progressed. On the vessel that carried him to a surprising future, Augustine must have looked in the direction of Rome and thought of his days in Milan, probably content with having left behind the torments of the flesh and the anxious chase for social prestige and worldly ambitions. The hull of the ship cut through the waters of the sea and moved against the embroidery of the tide.

Part 2
Ancestral games

Chapter 4
Origin

Despite being considered the solid ground of communication studies, or better because of it, the concept of culture needs radical reformulation. The antiquated perspective of culture defined it as an idealized abstraction, as a means of distinguishing human beings and the variety of social groups, as well as an edge severing humanity from the rest of the natural world. In that sense, culture is conceived as a set of immanent rules of social integration, whose purpose is to separate what is *ours* from what is *other*. The distinctions are neat: social groups are different because they have different cultures, and culture draws a line disengaging human beings from both animals and natural life.

These assumptions are untenable. The separation of cultures in the world of global communication simply does not hold anymore. Nowadays, no culture can aspire to isolation. Cultural identification is more than ever under the global pressure of information exchange. And why should we accept uncritically the opposition between what is natural and what is cultural, between what is given by genetic inheritance and what is acquired through human interaction? The separation of nature and culture stems from an inaccurate assumption that the natural world and the human mind belong to universes without bridges. It implies that the natural world is a realm ruled by blind necessity, resulting from the mechanical properties of matter and determined by unchanging links of cause and effect, inasmuch as human life and social interaction are free. This assumption is based on the premise that mind and matter are isolated dualities; but the dualism of nature and culture, of mind and matter, merely reflects an archaic metaphysical polarity: the one between body and soul. As we saw in the previous chapter, Augustine of Hippo formulated in theological terms the theoretical principles over which this tradition would come to rest.

4.1. The anthropological ideology

How can we continue to believe that human cultures are radically autonomous from each other, when even anthropologists embracing this conception of culture inevitably return from their field research announcing that the studied group is a meaningful whole, and that it can be understood?

Anthropological interpretations may be right or wrong, but anthropologists consistently offer an interpretation of social gatherings.

Yet, anthropology disseminated the ideology of culture as strictly conventional, stating that patterns of thought form an invisible but potent barrier that grants collective and excluding identity to social actors. Culture then becomes a set of rules antedating social actions – the cause of relative social determination, a stable grid resisting change; but why did it seem reasonable to understand culture as fixed frame acting over social life making it resistant to change?

At the foundation of anthropology in British universities, anthropologists were incapable of dealing with a pressing social problem that was very near to them: the destruction of the urban working class by the Industrial Revolution. And, without inclination or desire to idealize the English peasantry, they transferred their expectation of domesticity and social equilibrium – so inbred in the upper crust from where the anthropologists were recruited – to primitive social groups subjugated to British colonial rule. It is an anthropologist (Leach 1984: 4) who recognizes that his discipline suffered, since its beginning, from this kind of arcadian illusion. In any case, anthropological thought is mixed with a superior attitude that, in a condescending manner, identifies what is its *other*.

Even for authors (Gass 1989: 189) writing from the fringes of contemporary social thought, culture is seen as a term that cannot be easily stripped from prejudice and bad consciousness. How can we forget that the apparently neutral anthropological approach to societies branches out into two trends superficially contradictory that are, deep inside, complementary?[1]

On the one hand, in classrooms, anthropology claimed to be the defender of the legitimate humanity of the people it studied; but, on the other hand, it furnished the indispensable knowledge for the transformation of natives of colonized societies into servants that would be bought with beads. Anthropology, whose birth is a consequence of the colonialist legacy, presented a bad deal for the natives. In exchange for their symbolic validation as human beings, the colonized would receive a religion that was foreign to their traditions; at the same time, the integrity of their society was usually destroyed. Colonized societies would never be the same; they became part of the general plan of humankind, to which they contributed unconsciously and at their own expense.

The idea of cultural singularity was marginal in early cultural theory. The set of values prevalent in a particular social group was seen as part of a unique whole organizing the apparent diversity of humanity. The lines separating human groups were less important than the recognition of a common legacy underlining the multiplicity of cultures. The influential project of a cultural and anthropological theory of Edward B. Tylor (1871) was a typical nineteenth century pseudo-evolutionary scheme[2] that had its foundations in the Enlightenment conception of a progressive development of humankind. This would change, but, at the beginning of the anthropological research, universals mattered much more than particulars. During the nineteenth century, the encounters of cultural theorists with actual and singular cultures were indirect. Cultures different from one's own were not known from lived experience. This is not to say that early cultural theory had a sloppy or careless approach to particular cultures. Encountering a culture was a means of collecting data, never an end in itself. Collecting data served a higher purpose.

Anthropology started under a clear-cut division of labour. Government officials, missionaries, travellers, explorers, and other tentacles of colonialism would experience direct intercultural contact, leaving the processing and reflection of data to armchair theorists with deficient second-hand knowledge of actual cultures. Tylor, for instance, thought that his role was to deal with the conjectural aspect of anthropological imagination, fitting singularities into an overall design. Meanwhile, the actual contact with specific groups was left to missionaries already in the field, like Lorimer Fison and Robert Henry Codington, or to explorers such as Everard Im Thurn (Stocking 1983: 78). It is therefore no surprise that culture, the key notion of such an odd undertaking, could be regarded as suspicious, however careful the British Association of Anthropology may have been, requesting rigorous information, and demanding data to follow a meticulous prescription. Informers were ordered not to ask uncalled-for questions and, above all, to be precise (Stocking 1983: 72–73). Informers should follow systematically the script of *Notes and Queries on Anthropology*, blessed by the British Association of Anthropology.

The great transformation in cultural anthropology came when this division of labour was unravelled, at the onset of the First World War, after Malinowski's involuntary but deep immersion in the native life of the Trobriand Islands. Malinowski's experience would reveal the importance of living in the social group to be interpreted. The interpreter of a culture should learn to be a native; the anthropologist must be able to comprehend the native culture from the inside.

With the lived experience of dwelling among the members of a society, a completely new insight on culture emerged. To separate a fragment of a culture from the conditions of its use was deemed a hopeless distortion.

Nothing could repair it, not even the high purpose of inserting the segment in the evolutionary metamorphosis of humankind.

4.2. Cultural cohesion

In Malinowski's terms, the basic rule to understand society was to dispense with establishing the sequence of historical and comparative links between social groups. The main objective now was to provide a "functional analysis of culture" (Malinowski 1931: 624). Looking toward the past was replaced by consummate immersion in the present. The functionalist conception of culture maintained that social experience was an integral whole offering collective cohesion to social interaction. The task of anthropological analysis was to identify the cultural values and regulations in actual and occasional behavior (Stocking 1989: 98). Every incident in a group loomed from culture.

After Malinowski, cultural interpretation was anchored in contexts.[3] The major condition for cultural analysis was the observation and the actual interaction of the ethnographer with the interpreted society. Any other contention outside of the present context matured into a fuzzy supposition. Linguistic exchange, and by extension all exchange in a culture, was only "intelligible when placed within its *context of situation*" (Malinowski 1923: 306). The trace cutting across all cultural events and whatever communicative act that occurred in a social group was singularity.

The impact of Malinowski's recommendation was deeply felt throughout anthropology. More than anyone else, he emphasized that the effect of a spoken word is entirely dependent upon the context in which it is uttered (Leach 1957: 131). Linguistic meaning could not be grasped by philological and historical concepts. Malinowski's idea of a situational context as a criterion of meaning led the interpreter to link one lived stimulus to another, expanding the conditions for his understanding to encompass more than actual utterances. Language in context implied the system of culture, and correct interpretation of meaning "needed a detailed account of the culture of its users" (Malinowski 1923: 309).

Literal translation of an utterance, word by word, from one language to another, would never be sufficient to cope with the intricate question of meaning. If speech is "one of the principal modes of human action" (Malinowski 1923: 333), how could we do away with the cultural frame in which the action occurred, and which itself allowed the action to be performed? In *Argonauts of the Western Pacific*, Malinowski (1922: 459) declared, "no linguistic analysis can disclose the full meaning of a text without the knowledge of sociology, of the customs, the beliefs current in a given society."

Malinowski's pragmatic conception of language may have been innovative, but it was marred by his unwarranted conclusion that in utterance everything amounted to a mere reiteration of the very culture, the system of conventions, into which they occur. Linguistic exchange was then reduced to *phatic communication*. Language grew into the fulfilment of a social function.[4]

For Malinowski, cultural acts were the transformation of biological impulses. Culture had a tendentious practical nature. Kinship was the cultural answer to reproduction; cultural notions of hygiene were the cause of health (Piddington 1957: 35). To the functionalist interpretation, culture was a form of practical reason whose restricted goal was survival.

If that was the case, did biological needs generate culture or did culture create organic needs? The circularity of answers to easily interchangeable questions indicated that this was a false problem, a tautology haunting Malinowski's understanding of culture. But it is valuable to recognize in Malinowski's definition of culture the traces of a resilient idea of culture as a rupture with an original natural state severing humans from the natural world.

Anthropologists themselves criticized Malinowski's functionalist theory of culture. The most harrowing critique can be found in Claude Lévi-Strauss's *Totemisme, Aujourd'hui* (1962a) and *La Pensée Sauvage* (1962b). Lévi-Strauss argued against the disqualification of thought in archaic societies. The "savage" was quite capable of sophisticated and legitimate thinking. For Lévi-Strauss, all human languages can express the workings of abstractions, and moreover primitive classificatory systems are as sophisticated as the theoretical enterprise of modern science. The dissimilarity of human classificatory systems results from diverse social interests and intentions, without being considered a symptom or a judgement of greater or lesser intellectual achievement.

The logic of the modern engineer had to be distinct from primitive thought's *bricolage*. The achievements of primitive thought and science vary as an effect of how each culture approaches nature. The primitive mind pursues a strategy levelled at perception and imagination; science aims at reaching nature through the grasping of its structure, moving away from what is given to the senses. The *bricoleur* manipulates nature, handles what is at his or her disposal. The outcome of scientific inquiry is different. Nature is not a limit, nor a hedge, but an object to be treated in an altogether different light. As a product of scientific thought, the engineer is immersed in a world of concepts; his or her goal is to transcend natural events, trying to capture them conceptually, thus providing a discernment of their structure.

In Lévi-Strauss's terms, culture had a positive stand in itself, not to be reduced to a mere transformation of biological drives or needs. Culture could never be "an immense metaphor for reproduction or digestion" (Lévi-Strauss 1988: 28), as Malinowski suggested.

Social life followed models that were different from organic and biological interpretations. Nature and culture belonged to incongruent realms. The way to interpret cultures should be in strictly cultural terms. Thus, the appropriate interpretation of culture should be inspired, although cautiously, by linguistics. Lévi-Strauss endorsed a sociocentric – therefore conventionalist – conception of cultural and social life.

However, Lévi-Strauss's idea of culture kept much of what Malinowski defined as the traits of cultural experience. Culture was distinctively local and singular. It was an autonomous whole, prior to any behavior in a social group. Culture was a departure from the biological order. The duality of orders – one natural, the other cultural – in a state of mutual exclusion was clearly visible in both Malinowski's and Lévi-Strauss's notion of culture.

4.3. Nature approximately

The roots of the anthropological dogma separating nature and culture go way back to Augustine's intellectual legacy embodied in the thesis of Jean-Jacques Rousseau (1712–1778). Indeed anthropologists with methodological concerns have already clearly acknowledged Rousseau's influence over anthropology. When Marshall Sahlins (1977: 9) said that Rousseau is the true ancestor of anthropology, he was just reinforcing what Lévi-Strauss (1976: 33–43) had previously stated. For Lévi-Strauss, Rousseau was the originator of an attitude without which anthropology would be unthinkable.

Lévi-Strauss asserted that Rousseau did more than just adopt a dichotomy excluding nature from culture. In itself that would never be an original contribution, because the segregation of mind and matter, reason and emotion, soul and body has been a recurrent theme in the Western philosophical tradition. In the previous chapter, we saw how Augustine put forward a semiotic classification discriminating *signa naturalia* and *signa data*: signs that erupt in nature and the ones that are meaningfully traded among human beings, or – not to forget the theological dimension of Augustine's thought – the ones that convey God's meaningful will to humans.

The whole point of Rousseau was to affirm a synthesis that transcended the rift between human order and the natural world. Rousseau contended that understanding occurs only after the subject is free from his (or her) original position, and begins to think of his (or her) self as an *other*. Then, the subject goes beyond his (or her) intimate preconceptions. It is a cognitive attitude similar to the aims of anthropological understanding; prejudices should be abandoned in favour of a relativistic stance. The result is a sense of profound identity between me and the other, nature and culture, the sensible and the intelligible, humanity and life (Lévi-Strauss 1976: 43).

Forceful as Lévi-Strauss assessment of Rousseau may be, it is also clear that Lévi-Strauss sentimentalised the philosopher's view of the natural world. For Rousseau, the distinction between nature and culture remained untouched. Animals were radically alien to the basic principles of social life; they could never have enlightenment and freedom. According to Rousseau, only in civil societies, where human beings social only social actors can have the sense of compassion. Compassion would not be within the reach of animals. As with many thinkers who excluded nature from culture, Rousseau moved in circles; he talked of compassion as a condition of sociability, but it is in the midst of social interaction that compassion is possible. Human societies are a radical brake from the natural world. Therefore, in Rousseau, anything that is a part of the natural world was completely left aside. The natural world ignores any kind of social contract.

Civil society would have started from an act of territorial demarcation, yet fundamentally different from the disputes in animal conflicts. Someone may have declared, "This is mine," but society subsequently envisaged property as a legitimate possession. What was an individual and solitary act became, from then on, part of a socially shared convention. And convention will lead to the acceptance of mutual rights to all members of the group. Nothing like that could be accessible to animals. They would never grasp the notion of property, nor of any other aspect of society. Animals do not possess the necessary light to apprehend the proper laws of human groups. Likewise, Rousseau (1994), in *Du Contract Social* disqualified natural force as the ground of judicial order. Political power demanded another kind of authority, stemming from other shared conventions. Only conventions could legitimize authority. Eventually, even the most blatant despotism requires more than the exercise of force.

Rousseau's political theory, therefore, upheld that society is a pact, a convention born from other conventions. Social life is built to deny nature, an idea not very different from Augustine's rejection of natural drives. Societies cover nature with a dense net of contractual knots. To define social life as interconnected conventions also means to understand it as an exercise of freedom. This is another trait separating nature from culture. Culture is related to collective will and general representations, product of a social will; it is, therefore, the realm of freedom, while nature is repetitive, cyclical, the products of an unchanging order, the eternal expression of the same. The Augustinian themes are repeated in Rousseau, but under a secularised form.

4.3.1. Escaping Rousseau, chasing Darwin

Historically, Rousseau's thought helped establish the current consensus that freedom is the central mechanism of civil societies, but such commendable achievement is based on a flawed assumption on Rousseau's part. His claim that nature is the domain of necessity is a mistake, unsupported by today's biology. The natural world is neither fixed nor permanent, and it does not follow a prior plan or purpose set before the unfolding of natural events. Nature is under relentless change and evolution.

After the avalanche of Darwinian arguments, it is impossible to accept that nature is opposed to another realm of life, reserved and essential to human beings. Darwin's revolutionary understanding of biological processes leads to the recognition that in life "all past and present organic beings constitute one grand natural system, with group subordinate to group, and with extinct groups falling in between recent groups" (Darwin 1979: 450). Darwin's understanding of nature must be considered a radical departure from Rousseau's apprehension of natural life.

One of the reasons for the metaphysical postulate separating mind and matter to remain admissible for so many people is the massive presence in our culture of an essentialist theoretical vocabulary that has been faithfully kept by philosophy, theology, and even physics.

Moreover, the illusions of the essentialist tradition seem to be constantly confirmed by how Indo-European languages structure phrases around a subject that functions like a substratum.[5] The grammatical subjects are isolated and separate unities, under which the predicates of the sentence are gathered. The subjects suggest essences, while the predicates would be accidental attributes.[6] The subject *apple* is qualified by variable and non-essential attributes such as *green, sweet,* and *round*. Ordinary speech fosters the belief that there is a variety of permanent essences or kinds in the natural world. There should be as many essences as there are names functioning as subject of sentences, and, because God created the things of this world granting them names, each isolated name must indicate a separate set of beings, a species.

In 1831, when boarding the *H.M.S. Beagle* for a long trip that would take him during the next five years to South America, Charles Darwin, like many British naturalists, embraced the notion that species were created one by one and that they were not subject to change. Darwin then held a form of essentialist thinking, but in the trip to the Galapagos Islands, the assumption began to erode. He noticed empirical evidence that would contradict the idea of species as stable, natural kinds. Back in London, Darwin would present to the ornithologist John Gould a sample of mockingbirds collected in different islands of the Galapagos. Gould maintained that they were different species. Darwin was not convinced. The variation of the birds was the result of what came to be known as geographic speciation: the mutation of a species living in a separate niche from the one that the original species inhabited. Eventually, in one of his notebooks, started around 1837, Darwin (1987: 172) would remark that animals on separate islands ought to have become different if kept long enough (apart) with slightly different circumstances (notebook B, 7). Two facts about biological life were becoming clearer and clearer: there was evolution, and it was gradual (Mayr 1990: 19).

If early British anthropologists diverted their gaze to societies living at the edges of the destruction imposed by the Industrial Revolution, Darwin did just the opposite. As he explicitly stated, the dismal struggle for existence was a reality equally valid in the human and the natural worlds. In both realms, life is the result of a continuing substitution of solutions aimed at better and better adaptations, without which living organisms would perish. Humans and all other living beings exist at the centre of an arena where adaptive solutions would be enacted in confrontation with the environment. The adaptation of an organism to its niche determines its future. In the natural world, the species incapable of adapting to an environment under constant change is doomed to disappear. For all organisms, life is always a matter of either perishing or surviving; it is a frame limited by ceaseless births and deaths.

Facing the hard countenance of biological existence, Darwin concludes that life is always a delicate state, in which havoc and destruction should be expected. The insight comes from Malthus's gloomy prediction about the fate of industrial societies. Malthus (1888) saw the combination of geometric population growth with the steadiness and eventual shortage of housing and food as inevitably spelling disaster. Shortage of housing and food would increase prices, making survival a gamble in industrial societies.

Darwin picks up Malthus's prophecy and extends it to animal species that, incapable of either increasing the production of food or controlling their demography, are under the constant threat of disappearing.

In Darwinian terms, the inherent destruction in life processes receives the name of *natural selection*: in other words, the discarding, the preserving, and the multiplication of adaptive variations in the individual members of a species. The mechanism of natural selection delays evolutionary mutation, keeping organic solutions that have worked in the past and still function in the present, but disposing of them only in the case of change in the environmental conditions bequeathed to the individual members of the species. Natural selection has nothing to do with improvement. In fact, there is nothing in the idea of natural selection to suggest a cumulative progress (Williams 1966: 34). Individual organisms could not care less for the betterment of the species. They just want to survive. The quality of adaptive mutations is measured simply in terms of how long the organism survives. Reproduction and survival in life are the only objectives.

As life is marked by relentless destruction, the species procreates as much as possible. Under ideal natural conditions, with abundant food and without spatial restriction for multiplication, all species – whether plants or animals – can increase their numbers with each generation (Maynard Smith 1995: 43); but this is not what happens. Why are species limited in the number of their surviving individuals? The reason is that death before full development is the customary fate of many living organisms. Everywhere there is restriction of food sources as well as predation. Another species is just a likely food source. For Darwin (1979: 172), the tree of life covers the surface of the earth with dead bodies and broken twigs. To counter death, though, each new generation of organisms brings with it a new starting point in the gradual process of evolution. Two distinct individual organisms mate, and the offspring are altogether new organisms; diversity and variation are the cardinal manifestations of life. The evolutionary plasticity that sexual reproduction allows will increase the chances of survival of a species in a mutant environment, owing to a greater variation of its individual members.

The Darwinian conception of life had to reserve a great role for the individual. Individuals carry both the evolution and the adaptive improvement of the species. There are no essences, just individual organisms. A species is just a collection of individual organisms that may breed with one another. Darwinism shows that the arrow of mutation and evolution starts in individual variability and that the most precious trait of life is the variation of their populations. Apparently, with this notion, we find a real alternative to the prevalence of culture as a collective set of representations that precede and conduct human beings.

4.4. Predators and prey in interaction

The chain of living organisms is under the tension of mutual dependence, whose outcome is either survival or extinction. Of course destruction in nature occurs in many forms. It can spring from climate change, causing shortage of food in specific areas due to floods and droughts, or it can result from direct predatory interaction. Community in biological niches lived as mating, foraging or outright conflict. The natural world is always under steady competition. Open and generalized predatory pursuit predates any kind of cooperation and compassion.

Not even the mother and the foetus live in sublime cooperation or lofty compassion, sharing common interests. Since 1993, when Haig wrote his perceptive paper on genetic conflicts in human pregnancy, the picture is quite different. Harmony between mother and foetus is only an occasional and temporary equilibrium, and what defines their relationship is a conflict of non-harmonious interests. Mother and foetus compete for nutritional resources. Foetus's cells invade the inner lining of the uterus through the mother's weakest points: her vessels cannot defend themselves. The objective of the foetus is to manipulate the mother's physical condition to its advantage. The limit of this manipulation is the foetus's selfish interest: if she dies, the foetus will suffer dire consequences, and the cost of the manipulation will far outweigh the benefits.

The existence of compassion and cooperation is an after-effect, the aftermath of erasing or distorting a message that is inscribed biologically in every organism, most intensely in the sexual ones: "to explore the environment, including friends and relatives, to maximize our proliferation" (Williams 1992: 15).

At this point, it is necessary to present an explanatory model of the fundamental predator–prey link that informs the basic relationship of biological experience. Then, the question is how to describe predatory processes and what conclusions could be extracted for the understanding of human cultures and human society.

To fathom the mechanisms of predatory processes in the natural world is not simple: predation may be universal in biological life, but its concrete strategies are singular, local, and restricted to specific environments. A successful species in one niche can face extinction in another. As detached observers, human beings cannot know, with ease and without doubt, what goes on in the perceptual apparatus of a species preying on another; thus, how could one check any hypothesis about other species? That difficulty is aggravated by the fact that each species has its peculiar mode of perception, not necessarily akin to ours. Moreover, the act of singling out a prey demands more than strict perception. It requires knowledge and discrimination: the predator must distinguish food sources. How can we cope with such a dazzling variety of cognitive processes when the testimony of predators and preys cannot confirm or refute our hypothesis?

The grasping of predatory interaction emerges from the consideration of a capacity present across the natural world: the semiotic faculty of representation; in other words, the ability to produce signs. Not only in human societies, but also in nature one can see the proliferation of signs resulting from the interaction of living and inanimate beings. Nature is a semiotic theatre: just consider what happens when the shadows of twigs are projected on a wall: we have indices in direct causal relationship. Indexical signs are unstable. The alteration of the twigs' position changes the shape of the shadow or makes it disappear. The sign is completely dependent upon its model.

However, if one moves from the inanimate realm to the universe of living organisms, it is necessary to take into account that living beings can grasp as signs – therefore, as representations – what comes as indices to their perceptual systems. The natural world is duplicated. The receptor collects indices, which are reinterpreted as visual forms that entertain an analogical relation with what was an instant ago a physical event in nature.

In terms of the animal mind, the nervous system of a living organism designates iconically its position facing another organism, possibly its prey. If it is prey, the following will happen: the prey moves and triggers the whole process of capture. The prey then anticipates analogically what is moving in the environment (Thom 1983: 273). One can imagine how the process goes, through a flickering of analogical anticipations. The predator identifies with the prey, loses its analogical bond, for the prey is trying to flee, and concentrates on recapturing through forecasting what the prey would do next. Apart from pure and simple flight, the prey is left with two equally analogical alternatives: it can display mimicry and camouflage to deceive the predator or associate itself with other individuals of the species. Analogy is the most fundamental mechanism in biological life; it is the underlying quality of moving predators as well as the relatively stable environmental scenario where predator and prey get in contact with each other. In any circumstance, although digital in their structure, for both living organisms and environment, analogy drives the natural world.

4.4.1. Groups in the natural world

The most obvious mechanisms of defense in predatory interaction are mimecry[7] and camouflage, either through blending in with the environment or feigning ferocity. Owen (1982) mentions the case of the frog *Physalaemus netteri* that displays on its back the outline of big, bulging eyes; this allows the frog to point its behind in the direction of the predator, forcing retreat or abandonment of the predatory chase. Another example is the butterfly *Limentis archippus* that mimetically looks not like a predator, but a specific kind of prey, *Danaus plexippus*, whose taste is awful to predators. Besides strict mimicry, there is another option: the prey can associate with analogous individuals of its species with the purpose of protecting itself from predators. This is the propitious situation for the evolutionary origin of social behavior.

It is naïve to presume that conventions precede individuals who are then forced to adapt to the rules of group. In nature, where sexual replication prevails, absolute uniformity is not a given: any sense of permanence is unstable and subjected to change. If the most fundamental fact of biological life is individual genetic singularity – a topic to be fully developed in the chapters to come – the formation of a group is always precarious, and thus quite different from the ideal stability that anthropological functionalism used to postulate. All across the natural world, and this includes human cultures, the whole is a relationship made from the individual positions of the interactors; it does not simply follow a prior collective mold.

Undoubtedly, the natural world is speckled with groups. It could not be otherwise; natural life is overwhelmingly centered on the transference and the replication of genes within a population, and genetic replication is in itself somewhat a product of cooperation between individual organisms (Trivers 1985: 65). Looking at the gathering of individuals in natural surroundings one can recognize constant and uniform movements as if it were the orchestration of a force stronger than its parts. Could the regular formation of a school of fish – for example – provide evidence for the claim that a collective mold directly determines the action of each individual organism? If it were so, it would be tempting to give predominance to collective aggregations, therefore suggesting that individuals subsume their selfish interest to the group. Williams (1964 and 1996) argues that the collective mold of a school of fish evolves from the serial sum of individual behaviors. What seems to be coordination is really an illusion of order, quite similar to the impression one gets when seeing the photograph of a tight crowd trying to escape from a room aflame. From afar, the group appears to behave orderly; but, in the middle of the crowd, the experience is chaotic: the individual running in the direction of an exit door initiates a response in other individuals who would otherwise run in another direction or would remain motionless. A new response redirects the movement of the crowd. Group configuration is involuntarily drawn from individual responses.

Fear of predation determines the formation of groups. Inside a group, an individual organism reduces the possibility of being singled out by predators. Zoologists have consistently noticed that animals that gather in open spaces, whether in plains where herds graze or in the vastness of waters, more commonly form groups. Fish inhabiting coral reefs will not form a group: they can hide inside the various crevices around them.[8] The reason for group formation is simple: the organism in the middle increases its chance of survival. The margins of the group must be avoided. The danger zone between predator and prey should be as large as possible. The organisms in the margins and nearer to the predators have more chance of being devoured.

Independent of Williams and dealing with what he calls the geometry of the selfish heard, W. D. Hamilton (1996: 229–252) comes to similar and complementary conclusions. For Williams and Hamilton, predatory interaction marks the evolution of social behavior. Hamilton imagines the behavior of a group of frogs living in a lake where a snake dwells. The snake always attacks at a certain moment of the day. Before the snake emerges from the water, the frogs will move to the edge of the lake, terrified of other predators living on dry land. The frogs run away from the natural habitat of the reptile, but they have nowhere to go; they forecast that the snake would eat the one that is closer to it; the snake will feed itself with the least expenditure of energy. The danger zone becomes so small that piling-up is the only solution. The frog in the bottom of the heap tries to jump to the top, distancing itself from the snake gliding across the surface of the lake. In open or closed spaces, group formation is inevitable if predation is such a strong possibility. The frogs would avoid grouping only if they could hide individually. In a school of fish or for panic-stricken frogs, the way out is to be involved analogically with a protective whole.

The individual fish defends itself calling attention to the group. It is quite true that in group formation some individuals will inevitably be food for predators, but it is also true that in the company of others, the possibility of death is reduced. The more individuals are placed in front of the predator, so much the better. This goes against the interpretation that sociability exists because the group is the central evolutionary unit. A group is formed to protect the individual, not the other way round. For that reason, school of fish and other gregarious formations became an adaptive strategy in biological evolution.

Just consider what happens if predation comes from the middle of a gathering, not from its extremes. The prey runs away from the center, with the purpose of enlarging the size of the danger zone that separates predator from prey. For instance, a herd of cattle is grazing when attacked by a lion coming not from the margin of the group, but from a hiding place in the grass. The group is dissolved (Hamilton 1996: 247). To avoid concentration, each individual runs in a different direction. The dissipative formation is a direct consequence of what type of predatory attack is unleashed. The group is never formed because of a feeling of compassion for the other: it results from a selfish intent of preservation.

4.5. Cooperation and conflict within species

Outright selfishness is not the sole type of interaction in nature. Groups are formed for the purpose of avoiding predation, but, in the course of evolution, the original selfish sentiment can be modified. There are cases of relationships, in animal groups that must be described as ones of mutual benefit. In such cases, selfishness is still a vital motive of grouping, but it is not any more its exclusive purpose.

Mutual benefit is a variation of selfish intent, but there is more than that. People can act in a selfless way because they will also benefit from their deeds: generous charitable contributions will enhance their status in the community, and will buy them a ticket to the Kingdom of Heaven. It pays off; it is an excellent deal: eternity is bought with a transient donation. The sacrifice for the offspring makes perfect sense: the bequeathed genes will live thereafter. Sociability comes easier to organisms that have genetic links. Social traits are developed in direct proportion to the degree of kinship between organisms. Being good to one's sibling, one is somewhat also good to oneself: siblings carry common genes.

What is the explanation for collaborative actions between individuals that have no kin relationship? A pertinent example is the case of the little fish *Labroides dimidiatus*, which cleans up the mouths of bigger ones, and, in return, receives protection from them in the event of a possible predatory attack. Trivers (1971) postulates that in examples such as this cooperation occurs after many interactions between specific organisms. Both parties have something to gain in this situation. Trivers named the phenomenon *reciprocal altruism*. Slowly, and if reciprocal altruism is successful, the number of altruistic organisms in a population will increase. Cooperation and trust are learned and reached; they are not an original legacy of natural life.

Then what to make of the mutual restraint of animals involved in fighting for territory and mates? In an animal conflict between males of the same species, it is easy to see that the contenders are to some extent cooperating with one another; they seem to agree to avoid completely or treacherously aggressive combat; they appear to be involved in some kind of conventional or ritualized dispute. Why is it that animals – allegedly thoughtless individuals – are involved in conflicts that follow rules and ritualised tactics? Is it not better to employ any strategy, as long as victory is attained? The fact is that snakes fight without using their fangs; deer interlock antlers but do not hurt each other; fish grasp each other's jaws; and the fight is then a sequence of pulling and pushing; antelopes' combats are enacted in a restrictive posture with their knees down. And how about threat displays where physical contact does not occur? Who benefits from restrained behavior?

It would seem that the group is the direct beneficiary, but this explanation, like all that give credence to groups in natural selection, is blemished. Darwinian interpretation asserts that selection tends to occur primordially at the level of the individual organism and its genes, not only because each individual is a new and unique evolutionary starting point, but also because it is much more economical for natural selection to weed out particular organisms that cannot cope with an antagonistic environment. A single selective death should be enough to do away with the harmful mutation, and if the mutation is individually beneficial but harmful to the group, it will spread itself through the population, demanding its whole extinction (Maynard Smith 1972: 11). Individual selection is frequently stronger than group selection.

With this in mind Maynard Smith (1972, 1982; see also Maynard Smith and Price 1973) set out to prove that a "limited war" strategy is beneficial to individuals fighting, and to demonstrate that restrained contact develops into a preferred evolutionary strategy in the natural world.

Consider what happens in the case of three possible strategies to be developed in a conflict. The strategies could be *conventional tactics*, *threat displays*, and *escalating fight*. Conventionalized conflict and threat displays have in common the purpose of avoiding injury through the restriction of physical contact. Escalating fight, on the other hand, would lead to possible injuries. If both contenders employ escalating fight, the benefits of acquiring reproductive success (females are the fundamental motive of dispute in nature) are impaired by the possibility of extermination or serious debilitating harm. Reproductive success is certainly desirable but not at the expense of individual physical integrity. It is possible to combine the basic strategies; it is evolutionarily more effective to be involved in an initial conventionalized fighting, escalating only if the opponent does so.

4.6. Signs displayed

Suppose that the living organism can take advantage of a strategy whose cost is not any form of physical harm but merely the consumption of time and energy. That would be the optimal strategy. Evolutionary preference should be given to ritualized fighting and threat displays. Communication – the exchange of signs – would take place instead of physical interaction, being beneficial to both contenders. A belligerent interaction is transformed into a situation to be defined by communication. The initial function of all movements of a fight is altered: actions that were exclusively part of fighting mature into a message.

Even in the case of a strategy beginning with ritualized conflict that escalates to total fight, we would have a communicative scenario. Organism *A* starts as a ritual, in other words, sends a message to its opponent, organism *B*. If *B* accepts the message, then the fight will be ritual; but suppose that organism *B* either distrusts or does not accept the message and escalates fighting. Organism *A* reacts by escalating, following the received message. Here, we have more than mere conflict. Traded messages determine the kind and scope of animal conflicts.

Threat displays are even more advantageous to the contenders. It is quite logical that so many animals adopt threatening instead of total fighting. Animals assume aggressive postures; they emit sound, grind their teeth, but there are no physical fights. At some point in the conflict, one of the fighters admits defeat, the other wins, but none of them is physically hurt. Maynard Smith (1972) argues that in such ritualized conflicts both sides lose if the dispute goes on endlessly: precious time is wasted that could be devoted to other important activities. Therefore, there is a point at which the persistence in fighting is a loss for both contenders. When must displayed threats end? Is there a contractual rule binding contestants in ritualized conflicts? As the two fighters cannot have the same physical endurance, owing to distinct genetic legacy or even acquired skills, ritualized displays trick the opponent to its limit.

At first, the objective of threat displays is manipulation and deception. Even weaker animals will try to show themselves as stronger then they are.

Signs and behavior that they are supposed to represent are in close analogical relationship. A tenuous line separates true information about an attack and manipulative deception of an opponent. Organism *A* must not reveal its limit of endurance; therefore, the emission of a threat signal must be kept at its highest intensity for the longest period. The first to diminish the intensity of the threat is in fact admitting defeat and thus inviting attack. From this moment on, it does not pay to persist in the conflict. It is better to leave the dispute.

Viewing the fight as an interaction, where organisms *A* and *B* interact on equal terms, we have to recognize that deception is an occasional strategy. The repetition of deceptive strategies will leave the cheater open to codification on the part of the other contender.

4.7. A natural typology of human societies

To adopt an evolutionary perspective on culture brings with it significant advantages. It overlaps with some of the anthropological claims about human societies, and yet redefines human cultures, integrating them with biological life.

In this chapter, it was argued that there are two mechanisms in animal grouping, depending on how predation occurs. If predation comes from the outside, the ideal solution is to form a tight group; but when the attack comes from inside the group, individual organisms choose to disperse the totality, running each one in a different direction. It is quite the same in human societies. Human groups are organized according to two basic morphologies that ripple across humanity, and recombine themselves in various singular cultures. The singularity of cultural solutions is preceded by a morphology affecting the creation of specific cultural products and messages. In other words, two universal forms underline specific and particular cultural formations. The two basic morphologies are supported and created by "the circulation of complexity, of information, through the social body" (Thom 1975: 315), in short, by communicating signs. That is a phenomenon not at all different from the definition of fighting strategies in animal conflicts whose outcomes result from what messages are traded during the interaction.

Superficially, it would seem that the two morphologies are disparate and non-congruent ways of ordering the relationships of individuals and their collection, between the isolated parts and their whole. However, the fact that the morphologies comprise two ways of defining the relationships between individuals is a telling indication of the dominance of individuality not only among animals but also in human beings, underscoring the Darwinian intuition that self-preservation of individual forms of life is an active and primary force at the core of all life processes. We just have to remember the ruthless action of the foetus over its mother: complete and absolute exhaustion of the mother's nutrient resources would lead to her death. The foetus's selfish interest must be curbed, not just for the mother's sake, but also because it will be destructive to the offspring too: her death means its death. In human pregnancy, the alternation of biochemical reactions between mother and foetus can take care of the problem. Outside of the womb, humans create social rules that are shared and must be learned to allow effective interaction.

Human societies are consistently organized according to two major structure types with their own general dominant rules. The morphologies are complementary ways of defining the role of individual actors in society. One type is based on the subordination of the individual to the group, as it was argued in Augustine's description of the city of God, predominantly hierarchical and holistic. The other type is fluid: in it, each individual relates to another on an equal basis; its dominant principle and value is the autonomy of the social actors; social aggregation is an effect of several individuals interacting incessantly.

4.7.1. Hierarchical societies

In hierarchical and holistic morphologies, the whole defines the parts and the prevalent rule is that each individual should have a defined and preestablished place in society. Individuals are expected to occupy a position in a social pattern that is seemingly stable and that tends to elude change. In the natural world, flocks of birds and beehives are products of a holistic morphology. Among humans, such structured type of subordination is present in caste formations.

Bearing in mind that cultures provide the rules that mirror morphology, and examining traditional Indian culture, we can see an example of an extremely hierarchical and holistic cultural system. In it, one element embraces the other in a relationship of encompassing and encompassed (Dumont 1967; Khare 1971). The individual social actor does not construct a cultural identity from scratch. Individuals receive their identity from social personae that antedate and specify possibilities for life. In traditional Indian society, the form of the social frame is like a cusp. At the top of the social gathering, we find Brahmins, or priests, and going down the social slope, we find "below then the Kshatryia, or warriors, and then the Vaishyas, in modern usage merely merchants, and finally the Shudras, the servants or have-nots" (Dumont 1970: 67). Besides these four social categories, Dumont also identifies a fifth category, composed of the Untouchables, which is outside of the classification. Brahmins and Untouchables are as opposed to one another as purity is to impurity, as high is to low.

The Indian order of castes is a complex cultural system with restrictions placed on food, sex, and rituals. Social order is maintained through the communication of ideas concerning the purity of its members, and through kinship ties that guide endogamous connections, creating binds that anchor "the Hindu to his place in society, and curb the desire to strike on his own" (Yalman 1969: 125). Therefore, to secure such order, Indian society generates cultural signs that emphasize the message of social subordination. Culture creates a constellation of signs and messages powerful enough to act upon individuals. In the case of Indian society, the basic idea of subordinate contrast comes from the distinction between purity and impurity that culminates hierarchically in the figure of the Brahmin priest.

4.7.2. Individualistic societies

The other possible morphology would inevitably invert the idea of hierarchical subordination. In nature, the counterpart is the cloud of mosquitoes (Thom 1975: 319), whose movements are oriented from the point of view of the individual organisms interacting in the cloud. The glimpse of other individuals in the swarm corrects the course of the pattern and its constant possibility of disintegration. No totality governs the pattern; the pattern happens. Each individual is linked to another individual organism and all of them, through horizontal reciprocity, adjust themselves, avoiding disintegration. In human societies, clear legal definition of rules and norms respected by each individual can bar disorder.

Equality is the dominant rule in individualistic morphologies. Lévi-Strauss (1946: 643) would contend that, in the United States, the chief ideology is that "what is valuable for the part is equally valid for the whole." This does not mean that in egalitarian societies there are no rules.

With egalitarian relation as its rules, life in America, for instance, moves through a deep and ingrained desire for uniformity. The individualistic streak that his morphology fosters is reduced to a constant demand to conform, to become part of what is considered mainstream. The viewpoint of the common human being is deemed as valid as the extraordinary accomplishments of exceptional achievers.

In his trip to America during the 1830s to review its legal system, Alexis de Tocqueville (1994, vol. 1: 254–287) perceived this contradiction in United States society, pointing to the possibility of a tyranny of the majority. The action of an individualistic morphology is present in many areas of American social life. The First Amendment to the Constitution sanctifies the freedom of the press, preventing the government from creating laws that restrict free speech. It is amendment against the government, phrased to protect the individual, and to ensure the universal right of free access to individual consciousness. All over the social fabric, civil and contractual rights are extended to singular members of the community. Collective entitlements are under constant criticism, even if upheld. In terms, "fun," "pleasure," to be a "nice person," are ways of being cherished by the community as special individual, and are constantly prized in America. In this social setting, entertainment becomes a major industry. The discourse on the social role of individuals is centred more on rights than duties. Hierarchical morphologies are just the opposite: duties come first.

In societies where individualism prevails the purpose of social interaction is easily turned into the pursuit of individual happiness. The economic system is an end in itself, the market is a self-regulating and autonomous sphere (Polanyi 1968, 1975), exactly like the individual social actors. The economy exists to permit and to encourage individual accumulation of wealth. It is quite different from traditional societies that bridle economic to social interests.

4.8. Toward sex

Whoever complies with the anthropological concept of culture cannot go much further than identifying the cultural rules of a group. The rest, even dissent and social negotiation, is supposed to come without any other assumption or problem. The traditional concept of culture – with its blend of contract and conventions – draws a route to interpret cultural phenomena that moves from the group to the individual, ruling out the alternative that moves from the individual to the group. In cultural anthropology, the group takes precedence over anything: the individuals become secondary. The Greek idea of *paideia*, Plato's request of social coordination, and Augustine's idea of contract are the foundations of this idea of culture. What remains to be seen is if there is an alternative to the most basic foundation of Western social thought.

From the Darwinian conception of life, the view is different. The set of collective representations defining culture is just a model generalized from the perspective of individuals. Therefore, culture cannot bind them. At best, culture opens up individual possibilities. The conclusion is that the individual organism will use seemingly collective rules, subverting them, negotiating relationships, and doing whatever is necessary, even delivering honest and true signs, just to impose its self-centered interest. The tension between selfish interest and cultural norms is too valuable to be discarded.

As it was seen in this chapter, it is feasible to interpret human cultures without a rigid separation separating nature and culture. In the next chapter, the role of sex in the generation of individuals will be considered, taking into account how the exchange of signs emerges as the most primitive trait of sexual organisms.

Chapter 5
Sex, signals

The contemporary canonical conception of culture rests upon an ancestral continent: Its original legacy recedes as early as Thucydides, cuts through the discussion of Socrates and Hermogenes in the *Craytlus*, and is fully accomplished in Augustine's moral and contractual proclamation that all the created have an obligation with the Creator. After Augustine, a new concept of the social contract started. For Augustine, the contract was not a choice but a mandate: the created ought to live according to the norms of the divine Scriptures - the universal canon where one finds God's word – and where the eternal and unquestionable prescriptions for humankind are expressed.

Among ancient authors, this is the strongest possible idea of social contract; the covenant cannot be broken for it has been bequeathed through Divine grace. As robustly as nobody before him, and despite his strict theological nomenclature, Augustine initiated the trend that dominates until today, pronouncing that conventions must fully precede and determine not only the social and moral acts of individuals, but also their inmost choices, indeed their internal world. From Augustine onward, the contracts and the conventions were transformed into a synonym of communitarian life. The theological thrust of Augustine's formulation subsided, but still the rigidly contractual nature of social interaction persisted. Although secularised, conventionalist theories understood that cultures had to be inescapable, and their power was such that, without them, individuals faded away. In this tone, Louis Dumont (1983) would contend, piling analysis upon analysis of influential social thinkers, from Calvin to Hitler, that the individual was a social construction, an ideological effect, in itself nothing but a mirage. The rules of the contract were the exclusive social reality.

Nowadays, depending on how one sees it, this notion of culture is either a platitude or a timeless truth. Some could maintain that a proof of its theoretical validity was that it became silently an integral part of the conceptual apparatus of contemporary authors as diverse as Hegel, Marx, Freud, Durkheim, Saussure, Heidegger, Foucault, Kuhn, Feyerabend, the struturalists, the post-structuralists (Hirsh 1983) and probably whoever else might arrive without warning at the backdoor. It would be misrepresentation to deny that the conventionalist tradition has kept without fail an open-door policy to a wide variety of authors with incongruent objects as well as political and ideological agendas.

It is daunting to try to find a common thread in such a vast stretch of ideas. On an immediate level, conventions have been seen consistently as elements of integration. They would be the cement that took surprisingly divergent adaptive forms to countless social situations, varying within a social group and in relation to other human gatherings. Cultures could then grant societal cohesion, for they were agencies to coordinate and curb the potential attrition among individuals who otherwise would devote themselves to an everlasting and savage combat of all against all. At the same time that this appeasing dogma was constantly epeated, it was forgotten that the identification of conventions provided the instrument to pencil the discontinuous – albeit imaginary – line between natural and human history. Conventionalism gave birth to the image of a social life *in camera*. Nature and culture had to be hedged from each other at all costs.

The question now is to examine whether conventionalist theories are as solid as their tradition apparently suggests. The point is to consider the conjecture that rules and conventions are autonomous features without a biological basis, and thus restricted to human societies. If not, it would be indispensable to refuse the assumptions of conventionalist doctrines and, furthermore, to rethink the nature of rules in communication between individuals, not forgetting to address the social and political implication of this finding.

When conventions are the starting and the finishing points of semiotic analyses, the discussion leads to despondently circular arguments. Jean-Jacques Rousseau (1994) perceived clearly the logical snare of this kind of reasoning, and, immediately, he took care to shut down the gaping trap. Originally written in 1762, *Sur le Contract Social* established a firm border separating nature and conventions. Reason and will, conscience and liberty were the grounds of any social pact among humans. Rousseau found politically unacceptable the contentions of the proponents of social despotism who envisioned tyranny and social subjugation as facts of nature, based on presumably inherent rights of the strongest. He hinted that more than natural force supported despotism and tyranny. Despotism and tyranny depended upon political and thus contractual deliberations. For human interaction, conventions were everything.

In any case, society is a pact, the upshot of a contract, expressed and sustained through a system of working conventions. Endlessly, conventions are derived from other conventions and refer to other conventions. Human societies and conventions overlapped tightly and completely. Nonetheless, this should not be seen as the predominance and the triumph of conventionalist theories. The endless self-referral of conventions to conventions is just an indication of conventionalism's theoretical frailty. Quine (1969) pointed out the internal paradox of conventionalism: conventions are reached through an agreement akin to a board of syndics meeting around a table, using – to establish the contract – another system of conventions, indeed a language. This is a circular reasoning of interminable self-reference: conventions developing from earlier conventions that refer to further conventions producing more conventions. Conventionalist arguments propose a solution that reproduces the problem that it is supposed to solve.

5.1. The case for individuality

Considering biological life, conventionalism incurred a serious flaw. With no evidence to support it, conventionalism presumed that life ran along parallel tracks. One track housed the entire organic world, perhaps forking out in other branches, while another parallel set contained everything belonging to human life. That is perfectly consistent with the axiom that nature and culture should not mix. The error of this conception is evident in its direct confrontation with the basic Darwinian tenet that all organic life share the same root. When Darwin first proposed his revolutionary theory, the idea of common descent was a brilliant theoretical speculation, but nothing more than that. Today, however, it is obvious that Darwin was right all along: just consider the discovery that the DNA elements are a common molecular inheritance of all living beings. Life is unitary. It is thus nonsense to confer privilege to any species. The unity of the whole set of existing organisms comes from their ancestors; that is the rule of the physiological mechanics of either asexual or sexual life: life never springs from the projection of metaphysical and lofty essences.

If the references are to sexual beings, each generation of new organisms brings with it minimal variations that indicate probable new starting points in a gradual evolutionary process. According to Dawkins (1983), individual sexual organisms are both replicators and vehicles that do interact with one another. Individual organisms are vehicles of genes that change, wane, or survive because of interaction with other vehicles. Sexual procreation – the predominant mode of reproduction in natural life – requires interaction. Replicating sexual organisms must also be seen as interactors (Hull 1989). It is from sex that social groups are erected.

As either replicators or interactors, the individual vehicle and its genes are what matters, "never the type" (Mayr 1982: 46). There are no typical individuals. Average types or classes are nothing but subjective abstractions. While a purist may cringe at this thought, accusing it of reductionism, why is categorization into social classes and cultural identities not seen as the bluntest and poorest mode of reduction? If the individual is dismissed, collective principles dominate the proposed social interpretation, reducing the variety of social actors to a common denominator. What is this, if not reduction?

Culture, class, species, and any conceptual modes of collective sentimentality are incompatible with the grasp that natural life evolves. Because classes are abstractions, they have to be impervious to change. Where does change come from if all one has is the generalizing principle? With the collective mould, all variation is wiped out. It sounds awfully like the creationist wish: created separately, growing according to its own essence, each species is originally made as it looks like now. In *On the Origin of Species*, however, Darwin proposes a radical alternative, refuting the suggestion of a vertical and jolting progress: mutation in life is horizontal and gradual. Starting with the hypothesis that nature ought to be understood in its most immanent terms, therefore with something other than a projection of stable essences; Darwin's evolutionary conception underscores the processes of biological becoming and could never be reduced to the transformation of parallel and isolated basic forms. Life forms diverge from an origin and vary as time unfolds. For that reason, Lewontin (1983) called evolutionism a variational theory.

Darwin came to the recognition of evolutionary processes observing a paradox in natural life. In nature procreation is abundamt, and yet only a handful of organisms survive. Thus, in natural environments, there should be a perpetual sorting of possibilities, which could not be predetermined. It should be remember that the environment may change to the point of making it impossible for the individual organism to adopt one staple and overriding biological solution. The survival of the organisms may depend on various biological strategies. Individuals carrying the best possibility to counter or respond to environmental pressure will survive, thrive, and spread its strategy over a population. If any life has a fate, it comes from the complex interaction of individual organisms, their genes, and the environment. Changes in a set of organisms result from selection of types; but what survives is the individual organism that carries the better-suited biological solution for survival. Populational changes do not demand the prerequisite of accumulative mutation in the set of its members. Changes in a population are not the cause, but the posterior effect of individual mutations: species are those individuals that have to evolve independently of each other (Ghiselin 1974).

If homogeneity is a logical demand for the existence of classes and groups and that is not a given property of the natural world, then classes and groups cannot furnish the proper understanding of how natural life functions. The initial point must be the individual. Groups and classes may come later, and in many cases, if at all. The logic of life dispenses with explanatory notions grounded in collectivist assumptions. Life is startling uniqueness, the result of the mutual action of mutation and the weeding out of mutated alternatives, which allows for the differentiation of younger organisms in relation to their forbearers.

It is plainly wrong to state that the individual acts predominantly to perpetuate the group. Is the group everything? Is there nothing preceding the group? If the respondent is a conventionalist who deems human cultures as autonomous universes of rules, the answer is no. Yet, what benefits would the individual reap acting in self-sacrifice to assist an abstract and collective entity? Would it be just to contribute to the cohesion of the group? With no other purpose that would bring advantages to the individual, the idea is absurd. Suppose that the sacrifice is not absolute, still the problem never leaves. Take the case of an individual wasting personal time, involved in rituals, or making objects that circulate among other social actors. Why would that be done if there were no gains from it? There ought to be a better reason than the purpose of serving altruistically other social actors.

What seemed to be an explanatory dead end, and a desert of motives, alters completely if it is admitted that the individual has an imperative interest in the group. That is so, if the group is seen as a universe of sexual possibilities, a stage to act and exhibit qualities that can bring about the delight of being selected sexually. From this standpoint, rules do not exist to order, restrict, and control sexual activities; but to validate the competition for partners within the group.Only in relation to a patterned set of agreed upon rules can the choice of an individual be considered legitimate and excellent. The social and sexual awards thereby granted are fully merited. It is not that rules are projected over the parts that do not have the power to conduct themselves in social life. It is precisely the opposite: rules are systematically present in social situations to foster the good struggle for mates, not to repress competition, or to supplant sexual bids. The perspective of the individual is so strong that rules are frequently distorted and manipulated around the interests of social actors. The idea of a group unifying and organizing individuals is a theoretical idealization, a direct consequence of the sociological conjecture that cultures are impersonal techniques of coercion, conduction, and formation of social actors.

5.2. The case for sex

Although in practical terms human cultures do recognize the intrinsic centrality of sex for social life, building elaborate systems of marriage, mating, and kinship, cultural theorists have made every effort to dismiss its pressure. The general strategy is to separate biology from human society and the arguments are varied. Some of them are quite ingenious, like Claude Lévi-Strauss's (1945 and 1966) explanation for the phenomenon of the avunculate societies that relegate the biological father to a secondary role in the education and formation of the child; and as a result, the maternal uncle assumes the kind of social function that western societies reserve for fathers. In an avunculate society, the maternal uncle has a set of rights and duties toward the son of his sister, over whom he exercises social authority. An interpretation of this would be that societies have a special order that can dispense arbitrarily with the biology of human beings. The maternal uncle and the father are symbolic representations, indeed symbolic because conventional.

In a paper on the links between linguistics and anthropology, which is really an essay where Durkheim meets Saussure, Lévi-Strauss (1963) presented a remarkable theory, developed from the application of the structural phonology of Roman Jakobson (1991) and Nikolai Troubetzkoy (2001) to human kinship systems. The justification for the analogy with structural phonology is assorted: both language and kinship use discourse as their manifestation of surface; both are social facts; and facing both, social actors deem language and kinship as natural to a point of being phenomena that they do not even care to examine. The similarities do not stop here. Like languages use the same phonemes to compose words that will be made into sentences, kinship systems use kinship terms: *father*, *maternal uncle*, *mother*, and *child*. In the same way that languages are not only composed of isolated phonemes - and each language is defined by binary relationships of opposition between the phonemes - kinship systems reorder the relationships between the terms.

Inside one system *father* and *maternal uncle* may respectively occupy the poles reserved for respect and jocularity; in other cultural systems, the switch is total: the term *father* is reserved for the jocular role and the *maternal uncle* exerts the social position demanding respect: in that latter case, one is talking of an avunculate. A structural anthropologist would ask: Why do social actors have the tendency not to notice the arbitrariness of the relations between kinship terms? Lévi-Strauss (1966: 37) replies that it is to ensure group cohesion and social equilibrium. Social actors do not understand the nature of the connection between the terms, but they do have their defining structural oppositions clear in their minds so that they are never confused. The system is overwhelming and unconscious, but the specific kinship term is used and clearly perceived. In Lévi-Strauss' theory sex does not matter, what is important is the system – the generator of conventions - that acts upon the individuals conducting their expectations; individuals do not need to be totally aware of the kinship structure.

For the American anthropologist Marshall Sahlins (1977), the social phenomenon of the avunculate is not a strictly biological question. In his view, kinship is a pristine social phenomenon and has nothing to do with reproductive success. While Sahlins's accusation of theoretical naïveté is nothing short of astonishing, some implicit questions are left unanswered: what is the source of kinship systems? Are they conventional because they are social and social because they are conventional? How to get out of this classical circle? Why separate sociological interpretation and sexual selection? Where does mate selection happen if not in social settings? Nonetheless, Sahlins keeps on insisting that sex is not the mainspring of human behaviour. Kinship is therefore radically unconnected with sexual selection and substantially indifferent to reproductive success. In the path of Lévi-Strauss, Sahlins's views kinship as existing to guarantee the integrity of the group. Although with nothing to support the contention that the group is the beneficiary in human societies, Sahlins presumes that the perfect state of a social whole is one of permanent equilibrium. He could even admit that perhaps there is no culture without biology, but biology cannot specify the cultural properties of human behaviour. Moreover, biology in itself is cultural. A decisive example of this tenet would be Darwin's theory that reproduced unconsciously the social principles of Victorian individualism. Sahlins's claims are nothing but vulgar sociology of knowledge, and in a way as amusing as declaring that, because he wore English glasses, Darwin could never see the complete picture of natural life. The implication is that nobody could ever understand nature, for all biologists have been raised in human groups and refer inevitably to their social learning. Moreover, Sahlins seems to insinuate that evolutionary theory is the equivalent of just making baboons wear tuxedoes and puff the smoke of big cigars.

Another objection to sexual selection comes from British anthropologist, Edmund R. Leach[1]. He argued that the idea of sexual selection is useless for understanding human societies. Mate choice is restricted to only one type of kinship system, predominantly Western. Western kinship systems are constituted around the principle of generalized exchanged, in which some social actors are prohibited to be a mate – few in fact – whereas many are deemed free to be chosen as a stringent act of personal and individual will; but that is not the only kinship system. Most human societies never allow mate choice. A young man gets to the age of marrying, and then the group moves to choose the marrying partner for the young adult. It is the opposite of a wide choice. A system of restricted exchange complements the individualistic system of generalized exchange. In a restricted exchange system many are prohibited, whereas very few are allowed. The restriction can be such that the partner is already determined for young adults to marry, many times even before their birth. In the overwhelming majority of human societies, the notion of sexual selection would be inapplicable.

The three objections must be addressed if sex is to seen as an underpinning of human societies. Lévi-Strauss's intricate and smart theoretical elaboration rests on the premise that the father is *the* biological element in the avuncular relationship, whereas the maternal uncle is no more than symbolic, the effect of conventions, and thus purely social. Unfortunately, the premise is a misrepresentation. The avunculate does not dissolve sex into social relations, making it disappear. Quite the contrary, the avunculate lays even more stress on sex than one at first could imagine. To begin with, the mother's brother was also sired sexually. It is not as though the maternal uncle is purely social, as if that were even possible. In fact, concerning the degree of certainty of his/her sexual origin – a guarantee that the child will be properly attended because of stout kin relations – the maternal uncle is a better bet than the father. The maternal uncle is a more assured kin than any father can be. Someone else could well be the father, but nobody else could be the maternal uncle. All maternal conceptions are undeniable, while the paternal responsibility in a birth is naturally more or less hypothetical. The choice of the uncle assures that the whole edifice of social relations is not grounded upon shaky foundations. The avunculate is a sexual solution for the organization of social ties. Reproductive certainty is so critical for social relations that societies have come up with the avunculate system. In direct contrast to Sahlins's opinion, sex is the mainspring of social human behaviour.

Sahlins's sociological interpretation of Darwin's evolutionism is even more problematic. Sexual selection in Darwin's masterpiece of 1871 (*The Descent of Man; and Selection in Relation to Sex*) is not a male choice. Sexual selection is definitely a female attribution. In both natural and human arena, males display the signs that indicate their biological fitness, and then the females choose. That could never be an insight that would ingratiate Darwin with his fellow-Victorians. In fact, it is so against the mainstream that it took a long while to be even seriously considered, and still that is a difficult issue as Marlene Zuk (2000) demonstrated in her *Sexual Selection*. It must be remembered that another famous Victorian, Alfred Russell Wallace, the simultaneous formulator with Darwin of the evolutionist notion of natural selection, found the proposition quite hard to accept[2]. Darwin did not carry distorting Victorian glasses on the bridge of his nose.

Leach's critique relies on the assumption that kinship systems with no or restricted individual choice falsifies the principle of mate choice. Here, the confusion is between whether choice exists and whoever chooses. The fact that the young adult cannot choose freely is no indication that there is no choice or that choice is irrelevant. It is precisely the other way round: choice is so crucial that mature individuals make it on behalf of the young. Female choice is always threatening for the always insecure and disposable males. Ih many human cultures great energy is devoted to suffocate it, through the creation of a whole social apparatus, produced to bar what is a real possibility. If not, why would such cumbersome systems of cultural norms ever be shaped? Moreover, Leach's observation that love affairs in societies with restricted exchange of mates occur outside of marriages (as though it were different in societies with generalized exchange) only solidifies the bond between sex, choice, and society. The effort to segregate sex, choice, and society turns out to be quite pointless.

5.3. Live sex

Despite its extensive presence in the organic world, sex is a rather puzzling mechanism of procreation. Asexual reproduction would be certainly simpler, more economic, and less costly, therefore with greater chances to be implemented as an adaptive solution. Asexual replicators do not depend upon other replicators as sexual ones do. Sex is laborious and difficult. In the opposite direction of what the word reproduction does insinuate, sex is not strict reproduction, not of the whole organism anyway. If it were solely reproduction, sex would be like a copying machine. In such a device, each copy should be in practical terms absolutely like its model. Sex is rather different: out of two organisms, sex makes a third one that is not at all the same as the model, although made from the reproduction of its parts. Sex cannot be just reproduction.

Operative in a type of organic life so tiny that it cannot be seen without lens and microscopes – mainly in bacteria and single-celled organisms –

asexual reproduction splits one life form into identical others. The result of the splitting is a set of other organisms that are copies of the original begetter. What bacteria do is to feed themselves with genetic material before self-splitting and creating a uniform breed. Biological success does not come at all from the splitting itself, but from what it happens before the organisms are segmented, indeed from their genetic renewal; however, renewal in asexual organisms is rather limited if compared to what sex does.

Contrary to asexual reproduction, sexual procreation increases exponentially the recombination of the genetic material, avoiding the uniformity that asexual mechanisms foster. In the end, sex is a form of recombining the genes of the parents (in pairs) and grandparents (always in fours), and so on: the number of organisms involved in sexual procreation keeps on increasing as the genetic lineage recedes. Generally, sex has the function of matching recombination with outcrossing (Ridley 1994: 30). In comparison with asexual reproduction that does not beget diversity and singularity for it breeds uniform offspring, sex allows the proliferation of diversity in a geometrical scale, although gradually: not all sexual mutations are selected. Natural selection culls the mutated possibilities, picking the ones that better survive in a population of equally sexual organisms. With all trouble of its procreation, sex surprisingly managed to spread itself – victoriously – across nature.

The two types of individuals participating in a sexual mating are radically asymmetrical. One of them carries the reproductive eggs with all the nutrients – indeed in large sexual cells –essential for the development of the embryos. After the sexual fertilization occurs, this nurturing organism takes over the process up until the begotten organism is ready to act on its own in the natural world. The other mate in sexual procreation bequeaths the genetic material and, in the great majority of cases, little else; or putting it in a more balanced manner: it does little else when compared to the investment of the nurturing organism. An amazing feature of sexual procreation is that the nurturing organism provides so much and only spreads half of its genes. As any biology textbook proclaims: that is the consequence of the biological phenomenon of meiosis. As this is directly opposed to the genetic interest of females, why should they even procreate with only half of their genes, especially if the nutrients are all there in the nurturing organism? That is the critical part of the riddle at the core of the understanding of sexual procreation. Why would a selfish individual tolerate that kind of loss, when there is asexual reproduction that ensures the bequeathing of all the genetic material?[3]

From that viewpoint, sex is costly. Cost and benefit are the measuring standards for grasping the actions and choices of individuals that do take care of their own self-interest and therefore have a greater chance of surviving. The rule is simple: Whenever the cost outweighs the benefit, the organism either shies away from the proposed solution or – if duped or mistaken –wane and, eventually, it will be wiped out. From this position, sex is an apparent enigma. Females have steadily paid the cost of sex, and still sex remains a feature of the natural world.

Yet, that is not the only cost. Laggard and opportunist males will also pay a heavy cost for sex. This cost is a mutual burden for the nurturing organism and the giver of the genetic material. Recombination and outcrossing of the genetic material of the ancestors brings to the sexual offspring conflicting materials that create large and complex life forms subjected to physical decay and extermination, the inescapable senescence and death. The wage of sex is death. While sexual organisms face the certitude of death, life carries on in a process that goes back four billion years or more, and appears to be never-ending. Sexual beings hold a sentence in their bodies; the unknown is by what method or when the lethal assault will be performed. Sex breeds the strange paradox of the end that does not end, of a life always on the brink of finishing, but that will be continued through offspring that will go on procreating. Michod (1995) attributes this phenomenon to the mechanics of sex that renews and rejuvenates the DNA.

The wheel of life moves forward as it goes backwards. It starts again carrying the genetic legacy of the individual. That is different for an asexual organism that spreads itself through self-splitting. In a manner of speaking, the sexual being continues, although it does not. The vehicles of replicators are undone. New vehicles are multiplied carting old genetic material, now renewed. That is the only positive sparkle of eternity that sexual beings may have.

As if the cost of meiosis and death were not enough, females pay the cost of having to cope with males. While women do complain universally about the masculine tendency of being uncooperative, of not lowering the commode after using the toilet, of not being attentive to the baby, of not cleaning the house and of leaving underwear and socks around, much worse the carping would be if the partner were not a man, but a lion and its majestic mane. Lionesses do have reason to be grumpy living side by side with such extremely - almost totally - useless companions (Schaller 1972: 361) they lie all day under shady trees while the lionesses do the hunting, the feeding, and the care of the pride. If only the male participation in parental investment were the same as the female's contribution, some of the asexual reproduction splits one life form into identical others. The result of the splitting is a set of other organisms that are copies of the original begetter. What bacteria do is to feed themselves with genetic material before self-splitting and creating a uniform breed. Biological success does not come at all from the splitting itself, but from what it happens before the organisms are segmented, indeed from their genetic renewal; however, renewal in asexual organisms is rather limited if compared to what sex does.

Contrary to asexual reproduction, sexual procreation increases exponentially the recombination of the genetic material, avoiding the uniformity that asexual mechanisms foster. In the end, sex is a form of recombining the genes of the parents (in pairs) and grandparents (always in fours), and so on: the number of organisms involved in sexual procreation keeps on increasing as the genetic lineage recedes. Generally, sex has the function of matching recombination with outcrossing (Ridley 1994: 30). In comparison with asexual reproduction that does not beget diversity and singularity for it breeds uniform offspring, sex allows the proliferation of diversity in a geometrical scale, although gradually: not all sexual mutations are selected. Natural selection culls the mutated possibilities, picking the ones that better survive in a population of equally sexual organisms. With all trouble of its procreation, sex surprisingly managed to spread itself – victoriously – across nature.

The two types of individuals participating in a sexual mating are radically asymmetrical. One of them carries the reproductive eggs with all the nutrients – indeed in large sexual cells –essential for the development of the embryos. After the sexual fertilization occurs, this nurturing organism takes over the process up until the begotten organism is ready to act on its own in the natural world. The other mate in sexual procreation bequeaths the genetic material and, in the great majority of cases, little else; or putting it in a more balanced manner: it does little else when compared to the investment of the nurturing organism. An amazing feature of sexual procreation is that the nurturing organism provides so much and only spreads half of its genes. As any biology textbook proclaims: that is the consequence of the biological phenomenon of meiosis. As this is directly opposed to the genetic interest of females, why should they even procreate with only half of their genes, especially if the nutrients are all there in the nurturing organism? That is the critical part of the riddle at the core of the understanding of sexual procreation. Why would a selfish individual tolerate that kind of loss, when there is asexual reproduction that ensures the bequeathing of all the genetic material?[3]

much better than the price of the disaster of placing all the chips in one alternative that may not mature at all.

It is not difficult to see that the second alternative applies to not only aphids, grass, and rotifers in their widening of procreative alternatives but is also a metaphor for what sex does. Each sexual individual carries the seeds of a potential new beginning. From the logic of life as gaming, sex had to succeed somehow.

An objection to this train of thought is the remark that everything said about the lottery and gamblers is just an approximate metaphor, an inaccurate representation of life processes. The only contention valid in the lottery description would be that to diversify options is a better strategy in a game of chance and loss. It would be indispensable to produce more than a metaphor to support the case for sexual procreation generating countless individuals distinct from each other.

Life, however, is not just made of new beginnings. Life is mainly resistance to death. So the principles of sex should be effective in the maintenance of the health of individual organism. To consider the validity of the hypothesis, imagine a population of identical organisms. The identical organisms would come with the same reproductive capacity and the same immunocompetence. Without variety in this set of life forms, the organisms will be inevitably vulnerable to the predatory invasion of microbes and parasites, relentless sources of devastation of living organisms. If an organism were totally uniform, in the case of this predatory action the destruction would be complete and fast. Here the advantages of sexual procreation are clear: sex produces a population with a greater variety of capacities and strategies of immunization. Each sexual organism has a wider repertoire of genetic combinations. This should be an undisputed inference if sexual lineage is taken into account. Any sexual organism has two parents and four grandparents, and retroactively, each grandparent furnishes two more ancestors that will contribute to the enrichment of the genetic resistances of the original vehicle. Continuous accumulation of immunising resources is much more effective in the creation of resistances than random mutation. What other reason can there be for the adaptive success of sexual organisms spreading themselves all over nature?

5.4. The maintenance of sex: The fall of the virgin lesbians

While resistance to invasion from parasitical predators may explain the successful range of sexual procreation, sex has an odd feature that could provide the clue for grasping a major mystery of life. This strange feature is the recurrent 1:1 sex ratio in sexual populations. The 1:1 ratio is such an enduring equilibrium among sexual beings of different species that it has to be a structural property of sex itself. As Darwin did in relation to many other aspects of biological life, he (1998b: 255–263) was the first to notice the equal sex ratio in species as diverse as humans, horses, dogs, cattle, sheep, birds, fishes, and insects.

At the end of *The Descent of Man; and Selection in Relation to Sex*, after Darwin addresses the problem of sex ratio, one is left asking to whom the the strange 1:1 equilibrium would be be advantageous. Would it be good just for the group? The question is particularly relevant because, when talking of sex, it can easily be inferred that the individual could not be the privileged unit of sexual evolution. The impression is caused by the fact that the sex ratio takes into consideration the entire population of organisms, thus suggesting the possibility of being a group selection hypothesis. Furthermore, if the advantage of the 1:1 sex ratio were for the group, how is the tolerance and general indifference of sexual life forms to the mounting costs for the individual organism, specially in the case of females, explained? Females carry the physical burden not only of the pregnancy process, but also of nurturing the offspring, paying a high cost for procreating. Theoretically, sex may seriously damage an important tenet of evolutionist explanations, thus leading to the demissal of the individual as the privileged unit of the process. The complexity of the process daunted Darwin, who stated: "[I]t is best to leave its solution for the future" (Darwin 1998b: 268).

Yet, one should examine the problem of sex ratio from the viewpoint of its process.[4] While it is quite puzzling why life has not favoured three, four, or even more sexes, in an attempt to simply the problem, this book's argument will tackle only the duality of types of procreative organisms and evaluate the benefits of the 1:1 ratio.

The two asymmetric types – male and female - could interact potentially in four combinations: male/male, male/female, female/male, and female/female. Obviously the combination male/female and female/male are not different, but the same. Thus, the four combinations should be logically reduced to three: male/male, male/female, and female/female. However, before proceeding with the argument, one must bear in mind that this is not a reference to the use of sex as a means to cement and reinforce affection; therefore, homosexual or heterosexual love and modes of personal fondness are not the topics here. The issue is procreation. In addition, without forgetting that although technological genetic advances may allow for wider procreation alternatives in the future, the intention of the argument in this chapter is just to see how sex could develop the 1:1 equilibrium under natural conditions.

In narrow reproductive terms, the combination male/male has to be dismissed on an important ground. The cheapness and the frailty of male genetic materials would make the combination impossible without the introduction of a third female party. That would be a threesome, therefore, outside of the scope of the present argument. To understand the 1:1 sex ratio, it is indispensable to keep the interpretation along the austere line of the duality of sexes. If that is the case, one is left with the task of assessing the benefits of two options male/female, and female/female.

Although most of what will be developed in this section falls under the category of thought-experiment, indeed a theoretical speculation to assess not the reality of a fact, but to assess the limits of a potential explanation, it must stressed that the combination female/female is materially possible, for females carry all the necessary nutrients to breed in their cells. So parthogenesis – in other words, virgin birth, procreation without the contribution of the other sex, in fact with no male genetic material – is definitely an option in the natural world. The female/female combination must be examined attentively.

Imagine that it is possible to choose what children one can have. The question is what the best choice is. And more important, this choice would be made according to the parameters of self-interest, without any recourse to group conduction or social coordination from a centralized nucleus. The reply can only be that the best choice depends on a critical factor: the population of other living organisms. If the population is equally mixed, in other words, with roughly the same number of males and females, the choice is irrelevant. One might as well toss a coin to choose the sex of the offspring; but that would never be the best option if the population were composed of either all males or all females.

If reproductive success is measured by the number of offspring that one's lineage can bequeath to the pool of life, and if one is facing a population of all females, the choice is clear: one male would thrive in a population of all females. The rarity of the choice would grant greater value for the sex in minority, creating an incentive so that females would compete for the rare and precious male. On the other hand, if the population were of just males, then the sole female would be the most treasured organism in the set. Any small, though significant imbalance would prompt the process of valuing more the sex in minority: competition would be minimized and the procreation of offspring would be maximized for the sex with less presence in the population (Williams 1997: 53). For sex, being a minority is a valued asset.

Let one bear in mind that hypothetically in this population there is no meiosis so that the selfish criterion can be met. Without meiosis, the sole and valued male would selfishly beget males, and the female would produce females only. When introducing the time factor, what would happen? Time goes by, and because of the imaginary rule of procreation without meiosis, the ratio of the population would shift rapidly from a predominance of females to a dominance of males. Now, the situation is precisely the inversion of the initial one with all females about to be invaded by one male. The sole female invades the population of all males, and again, in the course of time, the population would tilt in the direction of a set composed of all females. The oscillation could go on indefinitely, bringing with it the perilous possibility of one of the sex types being exterminated. That is particularly serious for females. In the case of an absolute dominance of males, because of their material incapacity to breed, in end, the species – the whole population of organisms would go extinct. Females would perish because of the extreme success of males. Although a population of females would not automatically mean immediate extinction, the set could again be easily invaded by a mutant, and the dangerous oscillation would return, and with it the threat of extinction of females as in the previous situation of an all male population.

Consider the introduction of the variable "care of the offspring" to the problem. What is the best couple? Notice again that what is discussed here is a possible alternative; whether it could really occur is another matter. The options are male/female or female/female. The possibility male/male is excluded a priori, because males cannot procreate without the genetic material of a female, while parthogenesis is not a complete material impossibility. In a universe without meiosis, the female/female alternative is by far the best alternative. To the irritation of Biblical literalists, the ideal couple could never be as in Adam and Eve, or even Adam and Steve; it ought to be composed of females and virgins. It is tempting to call them provocatively virgin lesbians: virgins because they do have sexual procreation and lesbians because they are a couple.

Why is Phoebe and Eve – the virgin lesbians – the ideal couple? The females would bequeath all of their replicators to life's pond with the additional advantage of having a devoted companion to take care of the offspring. Although human males are much better at parental care than countless other males in the animal world, even to the point of some of them being excellent care providers, the overwhelming reality in nature is that the care of offspring is a female task. Who among men does not know that, after the birth of a child, a battalion of females – regimented among mothers, sisters, friends, etc – moves in to take over the house to help with the hard task of nurturing the baby? For a while, at least, husbands can be completely exiled from the care of their children. The general sense in the house is that husbands are incapable of even guessing what is best and most appropriate for the child.

A population of virgin lesbians should be a dominant biological solution across the natural world. While that is a material possibility in nature because of the phenomenon of parthogenesis, virgin lesbians were not favoured in the evolution of sexuality because of the dangers of oscillation in a uniform population. The immense reproductive success of virgin lesbians would progressively be upset by one random mutation; and, as said before, if too successful, the virgin lesbians would then disappear.

5.5. Winning without winning

Considering the selfish interest of the sexual organisms, the best reproductive strategy for both males and females is not the one in which either of the sexes dominates totally. Dominance implies the fragility of the triumphant ruler. The ideal situation for both types, the one that serves best the selfish interests of males and females is the ratio 1:1. There is no doubt that the 1:1 ratio is the most stable strategy for both males and females: its persistent equilibrium does not result from a master plan, but simply because males and females have no selfish reason to depart from it. To succeed fully becomes a problem as a result of the potential invasion of the sex type in minority. Both sexes can perish because of oscillation. Oscillation has to be avoided, and the only means for that to happen is through the maintenance of the 1:1 sex ratio. The presence and the maintenance of sex demand the ratio of frequency of sexes never to exceed the equal frequency.

An important consequence of this kind of reasoning is that there is no stipulation of design or plan for the natural world to have an emerging order. Order is not a prior condition for any kind of existence. Order surfaces as life evolves. It may sound counter-intuitive, but the simulacrum of a plan emerges from the mobbing of conflicting interests operative in a situation where each individual takes care of its own interest. The dominance of the 1:1 sexual ratio results from mutual self-interest of males and females. Using the cost and benefit criterion, it is easy to admit that females paying the cost of meiosis as well as being capable of tolerating even obnoxious males still gain a lot, for they avoid the harsh price of extinction. All benefit individually from the 1:1 ratio, whether procreators or progeny.

The arms race between males and females is not a unique competitive situation in the natural world. While the 1:1 sex ratio is a structural condition for maintenance of sex, sex seems to be also a decisive factor in another type of natural conflict: the relentless arms race between sexual beings and the opportunist parasites.

In a series of memorable papers, W. D. Hamilton (2001) stressed the contribution of sex in the race against parasites. Genetic recombination yields rare genotypes; and in such a diverse pool of individuals, with a greater variety of immunological systems, rare genotypes mean more resistances to the invasion of potentially devastating microorganisms. Even so, in the case of an ever-renewed parasite strike, many individual organisms perish, but not all of them. This is a priceless advantage of sex over asexual replication. Sex creates potential new lines of horizontal variations in the set of created organisms.

Paradoxically, the asymmetry between multicellular sexual beings and unicellular asexual parasites shapes the nest for the invasion of opportunist microbes (Hamilton 1980: 283). Sex is not just the solution that creates new resistances. The multicellular structure of sexual organism is also a source of frailty and problems for the begotten. Large multicellular organisms advertise many paths of assault and exploitation, otherwise unavailable if the life form were miniscule. Furthermore, cells in a multicellular organism have to be coupled with each other, weaving a web of proximity and transmission. While cells must cooperate with one another to ensure the concerted action and welfare of the organism, the urgency of cooperation also means a likely decrease in the awareness of strange bodies that might invade it. Each parasite could be a cell. That hints tragedy if the clear-cut distinction of cells and parasites is resolutely toned down. Many times, the host recognizes and reacts to the intrusive parasites; however, the warning signals may fail or come too late with dire consequences for the multicellular organism. If the organisms cannot easily and unequivocally activate its defensive system, the parasite might begin exploiting the host.

As Hamilton himself acknowledged, before him, J. B. S. Haldane had proposed that biochemical rarity is a possible guard against the action of parasites. What Hamilton did so brilliantly, and so consistently, was to point out sex as the evolutionary feature that propitiates constant innovation, without which whole sets of organisms would have perished faster. Hamilton (1980: 290) demonstrated that the greater the recombination the easier it is for sexual life forms to bar the invasion of the asexual ones.

Each sexual individual stores genes that may not be useful today, but will be kept for a future emergency. It is saved for the occasion when parasites might attack with a new strategy.

Other asymmetries exist between sexual and asexual life forms. Parasites are specialists and have a limited number of alternatives to draw from when the attack is launched. Sexual organisms are generalists, and the wider their range of capabilities, the greater chance of survival for a sexual offspring. Sexual beings thus hold a kind of resistance in reserve. For that reason, Hamilton (2001: 402) declared with precision "sex has to remain the essential back-up system of defence against disease." Hamilton's prescription is valid not only for the individual body, but also for superorganisms, like colonies, such as beehives. Back-up defensive resources must be available for prompt counter-attack if assailed by new parasites. Compared to the genotype of sexual beings, the asexual ones have the disadvantage of being much slower to return to previous genetic states that would match some of the ones harboured in the sexual lineage. Asexual organisms could only do that through mutations that would depend on external genetic materials, an uncontrollable factor that may or not may be available to them (Hamilton, Axelrod, and Tanese 1990: 3669).

In this scenario, sexual organisms ought to be concerned with their choices of procreative partners. Food may be a more immediate preoccupation and even a more universal one for most living organisms because in a natural world under the sway of selecting the best mate, many males simply never beget offspring. The set of existing males harbours not just dispensable organisms, but also utter procreative failures. The excessive number of begotten males is surely proof that nature is wasteful. Moreover, in an environment with males of varying biological fitness, and if mating is indeed the cause for having the definite set of defensive alternatives that ensures the survival of sexual organisms, the choice of the mate has to be a vital decision. No other lasting mode of survival is on hand to sexual beings. The individual organism can do nothing about the choices of its forbearers, and yet it is only too necessary to agonize and ponder about the present selection that will certainly affect not just the destiny of others, of its progeny, but also what will happen to a part of itself, namely its replicators.

5.6. Choosing a mate, selecting signs

Mobility and dispersal are the marks of sexual organisms. Larger sexual organisms with many cells can develop more easily localized systems of locomotion connected to other parts of the body; and increased mobility pays off because it allows the escape from predators, parasites, and diseases. Mobility and dispersal are crucial advantages that come from the evolution of sex. One sees that considering the case of grass and aphids. Sexual aphids have wings; asexual ones do not. Again, if the spread of grass is local, then the reproduction is asexual, but the sexual seeds of grass ride with the wind to grow into greener pastures. In a changing environment – and that means a habitat with rapidly changing parasites – sex is advantageous. Change favours sex; and, although incrementally, sex brings about change: "new combinations are perhaps all that sex can supply" (Hamilton 2001: 513). Contrary to that, individual parthogenetic organisms will be better preserved in unchanging environments with greater rates of equilibrium.The biological introduction of exponential mobility in the environment should inevitably lead to thoughts about the cycles of interaction between the organisms. Just like everything else in evolution, time should define the nature and the import of the cycles; nonetheless, either short or long cycles aregood enough for sexual organisms. The problem is the absence of change in which parthogenesis might arise. For the sexes, all kinds of movement are better than no movement at all (Hamilton 2001: 192), However, the duration of the cycle is an essential factor in the criteria of selecting a mate.

Organisms with shorter life cycles do not require fussy decisions. They are not going to last too long, so why worry? In this instance, it is quite reasonable to be indifferent or oblivious about the indicators signalling the health condition of the mate. If the cycle is longer, however, the female must *not* be indifferent as before. The development of elaborate sign displays in organic life is an effect of this logic. The longer the life cycle, the more factors have to weigh in the choice of a mate. In the case of a disastrous choice, the selecting female pays a lasting price, even more so if the offspring demands more investment in parental care. The choice has to be successful, and that depends directly on the female's capacity to discern the propensity or the resistance to diseases in the males. Elaborate rituals of courtship with daring and intricate movements are not irrational; they communicate precisely the health condition of the advertiser. The display announces: I am healthy, I could do this, I can give you great offspring: let's have sex. It is not, on the other hand, accurate to presume that the choice of a sexual mate is the strict matter of the physical conditions of the isolated displayer. Mates are selected in reference to the comparative qualities between qualifying individuals. It could not be otherwise: selection is always a verdict about the degree of biological excellence. Male competition is a dominant aspect in the final choice of a sexual partner.

Sex does not merely generate points of resistance to parasites. The choice of the partner is of utmost importance; and, as mentioned, that ought to be done considering the bulk of the organism's resistance against diseases.

An important problem now is to understand the functioning of the mechanism of sexual selection. The selector does not perceive directly the genes of the potential mate. What it perceives is a series of material signs from which should be deduced the biological fitness of the candidate. Brightness, colour, symmetry, size, formal and material qualities of skin, feathers, and other bodily components of the organisms do provide indices of resistance to parasites. That does not mean, however, that an animal has merely fended off diseases. It can indicate that the bearer has survived successfully the assault of debilitating parasites. And this is a good sign; indeed, being the outcome of a test, it is a better sign than the indication of the lack of contact with parasites. Hamilton (2001: 206) keenly observed: "The neatly patterned and coloured birds did indeed have a tendency to host rather neat and special parasites too." If a bird is larger and brighter, it has offered more paths of assail to parasites, and still has survived. That is surely an indication of its capacity of survival that will be possibly transmitted to his progeny. Colourful and symmetrically patterned birds are frequent choices for mating.

One must never forget, however, that the indicators of biological fitness are unstable and precarious in environments under the strain of mobility and change. Valid signals today may not be so in the future. Furthermore, signs of fitness can be brief, because the signals are never indicators of resistance in themselves; they have that meaning in direct relation to the action of mutant parasitical threats. For sexual organisms, communication must be constant and interminable, not only between individual organisms and the environment, but also among organisms living in the same habitat. Sexual organisms must develop a communicative capacity (although variable in its complexity from species to species) useful in both instances, in relation to the environment and in the interaction with other organisms.

Sex requires the development of systems of signalling with greater and greater complexity, for the obvious reason that the speed of the parasite sets the pace of the interaction between sexual organisms. As time unfolds, evolution comes to elaborate extremely sophisticated combinatory and self-sufficient systems of signs, specifically human languages. Even before the invention of languages, attributes like the shine of the skin and sturdiness of feathers or their patterns are made into signs. What was the causal effect in nature - an index is transformed into the full sign of physical welfare of the organism. In such a biological context, it is easier to grasp Charles S. Peirce's (1931, 1992, and 1998) definition of sign: something that stands for something else in some respect or capacity. At first vague and imprecise, Peirce's definition covers precisely the range of signs among sexual organisms. Any element of the material world can be a sign, as long as it carries out the basic condition of standing for something. Whether for animals or humans, the index is the most fundamental kind of signs; never the symbol like conventionalism has repeatedly claimed. It is not culture and its conventions that generate the communicative exchange of organisms: it is the command of sex.

5.7. Signs in a continuously drifting world

Conventionalist theories have narrowed the criterion of semiotic representation – something that stands for something else – to just signs that result from rules shared in a social (and frequently just human) community. It is easy to see that this parameter leaves out a vast array of representations. As though this was not enough, conventionalist semiotics makes a bold imperialistic move, declaring that all other signs are also rule-oriented. Generating rules are the clue to the understanding of the generated messages.

Conventionalism, therefore, helps perpetuate a distorted and segregating theory of signs, proposing that messages are composed of dissimilar elements – one, mental; the other, material – inextricably made into one. Messages are the effects of rules shared in a group. A specific conception of sign production comes to the foreground, discriminating any other type of sign as a mere after-effect of conventions.

In the case of sexual organisms selecting signs of health in potential mates, it is quite clear that the repetition of an attribute, after a successful choice, could be transformed into a stored knowledge akin to a convention – although inscribed in the organism's memory - and yet this only demonstrates that conventionalism discriminates the prior semiotic states that lead to codification. More important, how could the organism make the first choice if not considering the strict material composition of the displayed sign by the organism? Much before having a mental component, which is not a strict materiality, but an inference of a perceiving mind, signs are fundamentally material and causal. They do not need to be joined to a mental representation, though that can and eventually will happen in the course of evolutionary time. It is here that representations will surface into light; otherwise, they are not anything and from nothing; hence they are nothing.

A sign is not merely a thing, but a possibility emerging from the material interaction of objects of this world. And of all types of signs, the most basic because common to lower and higher life forms, from bacteria to human beings, are chemical indices (Wilson 1968).

Whatever their material manifestation, signs are always the possibility of something. From causal tracks, it is feasible to infer the indication of something else, like the biological excellence of the organisms bearing the signal. The quality of an organism can be expressed by many signals. They are represented in the brightness of birds' feathers, in Hamilton's (2001: 188-189) example, as well as tougher physical barriers, such as skins, membranes, and cuticles that would make more laborious the action of parasites. All that is a sign of endurance; and endurance is a measure of biological success.

Showy animals are more appealing not just to females, but also are more attractive to virulent microorganisms. Exuberant animals have run the risk, and thus carry indices that are living proof that the attack of parasites failed. Indices assert truths; they manifest actual existence, and contrary to pure possibility, exclude what was not the case. However, that could not be everything: the actuality of signs imposes a decision on what was initially possible, shedding away one alternative and selecting another, defining what was true in relation to what was false. When an organism selects an index, it acknowledges the indexical information conveyed in the signal. In the particular cases listed above, confluences of multiple signs – skins, membranes, cuticles, and chemical tracks and passwords indicate at first possible resistance to parasitical guests. The selector then makes a decision to adopt the signals and thus mates with the bearer. Consequences will ensue, and at this point, for the signal, that is another threshold: the signals are not causal and potentially non-recurrent tracks anymore; they have become signs that are more complex.

In a letter to William James, dated December 25th 1909, Charles S. Peirce clearly stated that, isolated or not, signs point to the possibility of something. Thus, any signalling situation must have a sign, a something, and a possibility mixed in such a way that one could presume: possibility will always be the prevailing characteristic of the modality of things. Therefore, each conceivable existence will trail along three modes of possibility: a may-be, a can-be, and a would-be (Peirce 1976: 868).

May-be, can-be, and would-be are different forms of insufficiency. None of them was actualised. A may-be presumes uncertain knowledge; a can-be indicates insufficient action; and a would-be depends on insufficient circumstances. Possibility presents itself as a more hypothetical abstraction in the case of the may-be; as an indicative and factual possibility pointed out in the can-be; and a would-be is comprised of a non-fulfilled conditionality. Conventionalism reduces representation to an impoverished would-be, a faked pre-determinism forgetful of all previous stages of possibility that are capable of enriching representations.

Framed this way, the modalities of reality are gradations of possibility, and thus the fullest form of existence ought to be continuity. Representations are the result and the interaction of may-be, can-be, and would-be. Expressed as would-be, as conditions for the generation of messages, conventions emerge always as the later stage in the process of up-and-coming possibilities.

If for a moment one dismisses the immense pressure of sexual selection, which, in any case, is a possibility not fulfilled for all male organisms – and concentrate on the foraging for food, the dominant scenario of sign-exchange is somewhat altered. Some organisms – the predators – forage for food; they need reliable information to launch their pursuit; while for others – the prey – it will be advantageous not to convey correct information, therefore increasing their chances of escaping the predator. Such conflict of interest implies the communication of two opposite types of messages: one, honest and the other, a deceptive possibility.

5.8. Deceptive and honest signalling

A prey moves in an environment. For it to be successful, it must not stand out against its background. Continuity between the organism and its habitat is a constant quality in the assortment of camouflage and mimic tactics leading to deception. Continuity is seen in the case of animals evolving to be optimally green and brown in vegetation where green and brown are dominant colours. Protection in the case of camouflage results from continuous fusion of organism to environment; mimicry comes from the evolution of organisms assuming, as they mutate sexually, the shapes and forms analogous to other animals. Individual protection is always the ultimate goal in delivering such signals. Thus, spotted markings in deer are concealing devices in places where "the sunlight falling through the leaves marks a dappled pattern of light and shadows in the undergrowth" (Owen 1982: 25). Deer display a visual pattern that resembles the sensorial quality of the background. A sudden movement, indeed the slightest motion, is discontinuous to the environment, and informs the predator of the presence of the prey: that may well mean a death sentence.

This is not to say that continuity is the sole tactic of survival. It is easy to identify cases where standing conspicuously against the background can be biologically advantageous. Many bright and easily distinguishable animals are also truly poisonous and distasteful. Predators notice the bright (aposematic) colours and may be attracted to them. After tasting the animals with bright colours, inexperienced predators are warned of their toxicity and distaste. The injury pays off because many bright and toxic animals have a great capacity to recover (Zahavi and Zahavi 1997: 9). The benefit of future avoidance on the part of disgusted predators outweighs the harm of the injury. False aposematic coloration has to be rarer than true toxicity; otherwise, predators would evolve in the direction of trying to devour colourful animals, despite their noxious taste.[5] Visual patterns can be conspicuous, deliberately, and overtly honest when the purpose is to mate or protect territory, its females, and offspring.

Sex is the breeding ground for the passage from competition to cooperation. Initially, the drive of sex sets one male against the other in a competition for females; it also fosters a competitive dispute between the chosen male and the females: males wish more sex than their partners do; for they can spread with relative negligence their genes among females that are inevitably choosy and selective. While the basic disposition of sex is to stir up competition, the competitive clash cannot linger indefinitely; it must stop and invoke in the end cooperation and trust. Trust requires truthful tactics other than self-centred deception.

Deception never disappeared from biological life because it is evolutionarily beneficial to organisms with conflicting interests. Either error can be a perceptual fault of the organism or it is provoked, as in the case of self-protection and stealthily moves from a predator to capture its prey. A perceptual error is biologically undesirable only if it brings evolutionary negative consequences, like the arrest of the spreading of the animal's genes through a population. Otherwise, errors and falsity are an integral part of natural interaction. Willey (1994) points to several cases where animals incite or are incited to make mistakes.

What is discrete and distinctive at close range may well degrade or have its characteristics attenuated in an actual habitat where signals compete with one another. Wiley (1994) mentions the unrelenting possibility of errors in animals, like frogs that mate in aggregation. The presence of a multiplicity of sounds distorts the frequency of the transmitted signal, therefore creating the conditions for mistakes. Another hurdle must be overcome: the organism should coordinate independent sources of information and reception, such as vision and judgement, matching the received sign to its interpretation. Errors can then occur, but as long as the error is not fatal to the perceiving organism, it is possible to abide with a fallible perceptual system. For that reason, fallibility was never purged from the natural world, remaining an integral part of it. The emission of false sensory clues to attract animals that would mistakenly transfer pollen, for example, would not be feasible without a structurally defective perceptual system on the part of the manipulated animal. The attracted animal that receives the signal cannot distinguish perfectly the model for which it is searching from the mimic that copies it.

To avoid the irruption of mistakes distorting the basic signal some conditions must be met. The signals ought to be conspicuous, for they compete with the environment and with other signals. That implies more than a mere natural trend to escalate signals. The cost of a signal is what qualifies it. If the cost is fully paid by the signaller, then the sign must be true; but that cannot be all. For a signal to stand out and be recognized by the receivers, it has to be relatively uniform and repetitious so that it can be a remembered pattern. Redundancy also allows less cost for the reception, although costly for the sender: a redundant signal can be guessed before its overall reception. The unevenness between the cost of sending the signal and the reduction of the cost for the sender underscores that what effectively matters is not the redundant signal but how much it reveals about the transmitter. A signal can be redundant, but the escalated intensity of its transmission is the decisively informative feature in the message. The important content of the transmission is if the sender is biologically fit to incur the costs of emitting such specific signals.

5.9. Why not deception everywhere?

Deception is, however, too contingent to be the sole trait of animal and human communication. Transmitters and receivers interact as autonomous players with conflicting interests. The receiver attempts to mind read the transmitter to reduce the effectiveness of deceptive signs that are not in its benefit. Therefore, this sparkle of sociability – two actors in direct relationship with each other – institutes the conditions to deter deception. Repeated and structurally redundant signals – so intent on calling attention upon the fitness and the capacity of the sender – lead to learning and to the identification of the signal and the signaller as cheats, making them ineffectual. Furthermore, taking into account that evolution presumes the repetition of strategies over long periods, manipulation can only be temporary and circumstantial. The manipulator and the manipulated are caught in a ceaseless evolutionary arms race. In the course of time, deceptive strategies are neutralized and stabilized. Deception is replaced by a mutual resignation that the best strategy is to trade reliable, non-deceitful signals.

For deceitful clues to function fully they would have to be intermittently renewed. In this case, animals should invent and create new signals that work – albeit temporarily – as deception; but the examination of animals trading messages shows that they prefer repetitive signals with a minimum degree of novelty. Wilson (1975: 200) in his comments on redundancy in animal communication describes the phenomenon declaring that animal displays in natural settings are rather tedious and predictive in extreme cases approaching the point of what seems to be inanity to the human observer. However, something appears to escape Wilson's observation: when the form of the signal is the same, it is easier to direct the attention to the person at the other end of the line. Knowing that the person answering the phone will always use the same expression or something very similar to hello, the receiver can focus on the individual features – tone, pitch, and rhythm of the voice of the speaker to identify who answered the call.

It is not only that animal signals tend to be repetitive. They use as signal what was before a movement that had another function. A present signal comes from an ancestral behavioural pattern; or in terms of Nikko Tinbergen's (1972-1973) notion: signals observe the *principle of derived activity*, in which, for example, a signalling movement of a bird descends from acts like preening, feather-settling, or movements in a fight or a contest (Krebs and Dawkins 1984: 381). The movement is a movement signal (Zahavi and Zahavi 1997: 67); and this occurs because of mere repetition, more akin to habit than to anything else. Novelty is not an aim, perhaps because it is not a primal capacity cultivated by the animal mind. This is an insight particularly puzzling if one bears in mind that, in the struggle for individual existence, deception seemed so logical and necessary and with it the requirement of new strategies between competitors that have mutually conflicting goals such as eating and avoiding being eaten.

Redundancy establishes a shared pattern between competitors, thus allowing a more reliable assessment of each performing organism. Crucial, in this case, are subtle differences of accomplishment rather than repetition *per se*. Redundancy is a condition for judging competitors that divide a levelled field. Another purpose of redundancy is insurance against error and misapprehension of a message. To test and enforce bonds in a group, ritualised and redundant signals gradually evolve into stereotypes (Zahavi 1980: 77). Stereotyping and standardization are evolutionary strategies that leave deception behind. They are not exclusive attributes of human cultures. They also occur regularly in the natural world: stereotypes and standardized signals show that animals are more committed to reliability and truth than to deception.

Truth, reliability, and honesty are such forceful products of evolution that even deceitful signals appear to depend and mature from an honest signal. A deceitful signal would have more chances of success when a sender takes advantage of a receiver's rule of behaviour (Willey 1983: 170). Its efficiency is the misapprehension of the receiver. In such cases, the deceiving signal can only be a sporadic strategy: its repetition will leave it open to codification on the part of the receiver, making it recognizable as a cheat in the future.

If evolution selects honesty, deceitful signals should become rare. The only alternative stratagem close to deception is the suppression of the signal (Cheney and Seyfarth 1991). The ensuing silence is the culmination and the limit of deception. Silence is the least noticeable of all deceitful strategies and, for that reason, the most efficient one. In evolutionary terms, truth is not a matter of a projection of an eternal, timeless Being, but what succeeds from becoming. This kind of truth follows from an individualistic viewpoint, caused by the counter-adaptations of organisms that have been subjected to manipulation.

5.10. Truth without conventions

Honest and true representations are straightforward effects of the cost and the benefit of emitting a signal. Contrary to what Plato defended and helped to institute as a philosophical boundary of veracity, truth does not have to be the projection of timeless Ideas, hovering above individuals who strive to reach it. Truth is related to the communicating organism. The Israeli biologists Amotz and Avishiga Zahavi have argued potently that truth is a biological necessity expressed as what they called the *handicap principle*. When first proposed, the handicap principle was received with scepticism[6]. The reaction was that it could not be the case. The handicap principle postulates that the emission of a signal involves much more than a potential deception. Honesty in communication is openly related to the signalling costs.

Signalling costs shape honesty and deception. Deceitful signallers would feel most the cost of their messages (Bradbury and Vehrencamp 1998: 663). They would pay unbearable costs in emitting signals that are beyond their capacity, sometimes at the expense of their own integrity.

"The larger the investment, the more reliable is the message"(Zahavi 1993: 227). In the end, the effort simply does not pay off, and deception is avoided. Only the organisms with highfitness would be able to deliver the message, for it would be too costly for a lowerfitness organism. The handicap principle has a definite impact over the understanding of how it is possible to transmit nonverbal, nonconventional signs that carry effective, in other words, credible information.

Zahavi and Zahavi's handicap principle explains the apparent paradox of waste and excess all over the natural world. This revolutionary idea is rather simple, and can be stated in the following way: if every communicated signal imposes a cost to the signaller, then a handicap is a warranty of its true biological excellence, a badge to be displayed for sexual selection. The costlier the badge, the more truthful it is, and the more it would be spread in populations. The handicap principle supplants narrow views of the Darwinian idea of adaptation. Looking at this phenomenon from the viewpoint of evolution theory, handicap signals are the outcome of their evolutionary development, a notion quite distinct from the dogma of mere utilitarian and adaptive selection. If adaptation is measured by the equilibrium linking organism and the environment, why is there abundance and inflation in the natural world?

Handicaps demonstrate the physical abundance and the biological superiority of communicators. An unscathed comb indicates that the cock has suffered no injuries, being a good fighter. It is therefore sensible to avoid confrontation with it. Despite being a physically unfavourable condition, a dashing and ostentatious peacock's tail is no hindrance to the survival of the bearer. A peacock's tail signals to a peahen that he will breed outstanding offspring. Large snouts in elephant seals block their visual field, and yet the fact that a mature mate carries a voluminous snout is an evidence of his fighting competence. Only the best and the strongest could afford to sustain such handicap.

Many other examples can be listed: birds and circus artists taking what looks like impossible leaps, the altruistic acts that aim at creating social awe and prestige not only for our millionaires, but also for birds, bees, and muskrats. One cannot help but think of Jimi Hendrix playing the guitar with his teeth, a chess grandmaster facing many contenders at the same time, and giving up his queen. Michael Jordan's flashy fights to dunk a baseball to get the same points as any other player in court, Charlie Parker drugged out of his mind, but still weaving wild improvisations, the Egyptian pyramids shining in the desert to advertise the might of the Pharaohs, the modern believer thumping on the Bible to affirm that the words in this book have to be taken literally: he not only believes like all other members of the congregation, he does it to an utmost extreme.

There is no reason to emit faked signals when the deceiver does not posses the corresponding biological quality to cope with the burden of the signal. An expensive signal can only be delivered if its truth content is not a fake. Honesty and truth are indicators of the biological fitness of the signaller. Over all, cheating is inefficient. High frequency of cheating would tend not to evolve. However, from the viewpoint of individual and opportunistic organisms involved with temporary conflicts, oblivious of the extensive unfolding of evolutionary time, the upshot cannot be the same. Weaker organisms may be tempted to show themselves stronger than they are. Cheating works punctually, although with greater difficulty in the long run. The reason for this is that signals and the behaviour they represent are in close analogical relationship. The intensity of the transmitted signal should exhibit the intention of putting up an intense fight. The decrease of the intensity of the signal implies its opposite: it hints at weakness. For cheaters, the effort may not pay off.

The next chapter starts from the recognition that the sender – whether animal or human – has the choice of adopting deceitful or honest signals. Apparently the pressure of evolution and social strictures should lead to honesty, but that is not what frequently takes place. How is the decision to deliver a deceitful or an honest signal made? Is it the result of coercion from an overpowering social whole? If that were so, there would be no deception in social life, and evolution would have done away with deceitful representations. Then, what are the conditions that determine the choice of one signal or another? Is it as the Zahavis have been stating that true, reliable, and honest signals are the inevitable outcomes of living interactively? How can it be so if countless deceitful deliveries of signs cross human daily life? Or is it feasible to take advantage of deceitful signs? If so, when? Evolution may favour honesty, but the actual players in social situations are just living the stark chances of the present, pursuing their selfish interests. What sign strategies are most appropriate for social life?

None of these questions can be answered without careful thought about the specific choice of strategies made by individual players. The final chapters of this book will evaluate the plans of actions that may determine the signals used and implemented in individual action. The option is in favour of comprehending what is known theoretically as a game.

Part 3
Individual games

Chapter 6
Strategies

If evolution were a straight arrow, deception would have been wiped out from animal and human societies; but it was not. The coexistence of deceptive and honest signaling is an upshot of the strict conditions that fostered the sending of misleading signs: in other words, the effect of the self-interest of the players and their awareness that perceptual systems are imperfect, thus propitious for the manipulation of other players.

In the interaction of two autonomous players, guided by the compulsory self-interest to survive, player *A* wants to take advantage of player *B*. Hence, it would be in *A*'s interest to deliver a deceitful message. The attempt may succeed, as it can happen if the deceiver consciously or not sends a signal close to one of *B*'s established patterns of behavior, or if the signal is so surprisingly and so bewilderingly new that *B* has no way to identify the cheat. *B* gets the sucker's payoff.

Now, it is only too reasonable for *A* to repeat the strategy that was successful in the first interaction. Immediately or not, repetition must be expected. In the case of no repeated interaction that would certainly mean that *B* has fled the niche. As shown in the last chapter, fleeing and mobility are acceptable reactions for sexual beings. Inevitably, another player will take the place that had been B's.

To simplify, one must assume that repetition occurs soon enough for player *A* to remember that its initial action was successful. In this setting, *A* does not have to have more than a sliver of recollection. Suppose then that the repeated action succeeds again. That means another repetition. If *A* is so successful that *B* never gets it, that implies *B*'s absolute exploitation and eventual extinction; but imagine that *B* counters *A's* action at some point in time. *A* will not be able to repeat the initial action that was successful in the past, and *B* will be able to identify and eventually store the knowledge of what was deceitful and what was honest. Honesty and deception demarcate each other mutually. That would be a structural reason for their persistent coexistence.

However, that is not all. Honesty emerges just from the same self-interest that gives rise to deception. Players committed – as they ought to be – to the preservation of their own integrity would inevitably fuel a relentless arms race. Player B would have to move closer to player A, and player A would have to upgrade the initial strategy to another level. It seems a never-ending process. The passage from deceptive signaling to its recognition as dishonest delivery is like the swinging of a pendulum.

6.1. Anatomy of the game

Nonetheless, one should consider whether rules and conventions can establish and determine – and in what way – the actions of individuals, and thus reduce the rate of unpredictability that the oscillation of truthful and deceitful messages produdes. Is it correct to assume that individual organisms follow rules because of coercion of the whole over the parts, or do they do it because their self-interest was not effaced? In many circumstances, it pays off to abide selfishly by rules. If so, the whole would be incapable of overriding the parts, and that would hint at an inference fully against what Plato and Durkheim in different ways stated about the social dominance of collective thought. Thus, when individuals act according to their self-interest, the idea that common rules are efficient enough to coordinate interaction - for they express the higher interest of a collective totality – ought to be logically invalidated.

Is there a social theory that does not annihilate competitive and self-centered individuals and still accounts for cultural rules? According to conventionalist assumptions, competition is an insurmountable obstacle that inevitably drifts social actors away from the cooperation that should be the characteristic of social interaction.

As any game starts with axiomatic conjectures, all the analyses here suppose that the players are individuals with the same capacity to discern what is in their own rational self-interest. Of course, when presuming that, the purpose is not to sustain that this is what happens all the time in daily interactions. To draft a theoretical model is to propose an idealized abstraction entertaining complex relations with reality. Model and reality cannot be confused. They must be understood in their dissimilarities. While models have a reach and degree of mental existence that cannot be altogether absorbed into actual social interactions, it is undeniable that - even if partially - theoretical models are present in social strategies. The theoretical model is a normative imposition over reality and never an inductive abstraction derived from concrete circumstances. The purpose in generating a model is to create a grid that can explain an array of typical situations. Therefore, a hypothetical model should explain more than one case, leading to the determination of the most effective rational strategy when there is conflict of interest. At least initially, the fundamental mode of interaction is always the one without convergent interests on the part of both players.

The starting point cannot be a rigid set of shared conventions. Otherwise, the interpretation will slide into the circularity that was criticized at the beginning of the previous chapter, "Sex, Signals." Here, any common pattern between players will be weak enough not to undermine self-interest and therefore rigidly coordinate mutual interaction. It is thoroughly a situation of conflict of interest.

In a game, the players are independent agents, each taking care of their own personal interests and thus deciding upon the best strategic plan to follow. Thus, the main concern here will be the evaluation of general strategies (i.e. complete plans for action), which come from the recognition that there are clashes of interests, which demand decisions about how to proceed to be successful in the game.

As both players are equally endowed, and therefore competent enough to second-guess that the opponent will act according to the same goal, each adopted strategy will necessarily have to take into account the possible movements of the opponent. As any game situation is strictly relational, the outcome of the game will depend not only on the actions of *A* but also on how *B* resists to *A*'s movement. Furthermore, the drama of a game develops from the conflict of players resisting the adversary's intention to invade its strategy as well from whatever comes from the outside of this frame of interaction: and that may interfere with the strategies of the individual players.

The players will make decisions that affect the others; and they will do that not as a consequence of their psychological constitution or in obedience to societal rules. The emotional and psychological make-up of the players would offset the initial axiom that they are acting in a completely leveled field. Moreover, the players are also equal optimizers in search of the best solution that fits their interests. The players should not have any relevant differences that would give a head start to one or another. Codified as social mechanisms, societal solutions are inefficient to determine the outcome of the game. The reason is obvious: societal solutions are known to both players and as such easily counteracted. The assumption is that the solution of the game has to be rational.

The dynamic interaction of equally endowed optimizers is a dominant trait of the game. The player will act according to what is supposed to be the best rational choice of strategy. The whole point is that the dynamics of the conflict will affect and shape rationality. Rationality, in this sense, does not refer to an abstract set of logical principles that determine the appropriate inferences. The players decide rationally what strategy to adopt in relation to the other player in the game. It is what Aumann (2000) calls individual rationality.

To understand that the interaction of the players follows a set of guidelines is as important as the recognition of the clash of interest itself. In this sense, a game is more than an outcome; it is an outcome resulting from conditions known as rules, and they are therefore mutual and reciprocal. It is through the observation of common norms that the outcome is deemed socially legitimate. Although incapable of settling the outcome, rules are active at the beginning and at the end of contests.

Because the game has rules, the players behave in the manner that they do. Think about American football, improperly named for the players predominantly hold a strange oval ball with their hands, and basketball. Why is it that in a football match the players are constantly bumping into one another and holding down the players of the opponent team, whereas physical contact is barred in basketball games? That occurs because the rules of each game prohibit a specific kind of action, simultaneously encouraging another, and therefore establishing a logical connection between the acts and their legitimacy. In football, toppling the opponent is legitimate; in basketball, one is not allowed to grab the adversary; and if that happens, punitive sanctions must be enforced. Rules establish equivalences. It is not the movement in itself that is valid. Validity comes from the relation of the act to the rules.

Consider now how points are scored in a game. In football, the player must run holding the ball, dodging the players of the other team that will defend their field against the invasion, tackling and trying to stop the invader. If the attacker manages to advance to the end of the field and puts both feet across the line, then the attacking team scores six points plus the right to kick the ball into the adversary's goal. If that happens, one more point is added to the six already scored. In a basketball court, a successful ball thrown in the hoop means two points, but from outside of the three-point mark, three points. Again, the meaning of the action is established by the set of rules common to all players in the game.

The existence of rules guiding the conduct of players indicates that rules must be *common knowledge*. Common knowledge is a notion established in David K. Lewis' *Convention* (1969), and this means that besides knowing the rules of the game and the goals of everyone, "each player must be aware that each player is aware that each player is aware, and so on"

(Aumann 2000: 81). These are basic conditions for the interaction of the players; both players must know the prevailing rules and conventions; and each player must be aware of the other's awareness. From their condition of common knowledge, rules guide the dispute between the players, but little more than that. Games must have more than rules; they must have a point; and this point is reached through the players' decisive choice of strategy.

Games, then, are composed of three basic elements: players, rules, and outcomes. In games, at least two players in a situation of conflict of interest interact following rules reciprocally known. The player's choice of strategy determines the outcome of the game. Thinking about outcomes of the game, one can have basic possible situations reflected from the axiomatic admission of conflict of interest: there will be victories and defeats. Relating the two basic situations of victory and defeat to the two basic players, *A* and *B*, one sees four outcomes: *A* wins/*B* loses; *A* loses/*B* wins; *A* wins/*B* wins; *A* loses/*B* loses; revealing the existence of two kinds of games: games of competition (win/lose; lose/win) and games of cooperation (win/win; lose/lose). Such is the basic categorization of games; some are competitive; others are cooperative.

6.2. Complex utility

However central it may be for the understanding of a game, the concept of conflict of interest must not be seen as expressed only by the fact that one player's victory is another player's defeat. In political games, a candidate may run not so much for the purpose of winning, but to make a political point. Despite his insistence that he was in the race to defeat the other candidates, in the last two American presidential elections, Ralph Nader's third party candidacy falls into this category with his aim being to highlight environmental issues and economic policies rather than to become president. Or contemplate a father who plays chess with his son, allowing the boy to win as an incentive for the child to continue playing the game.[1] The complex interests of the players determine the point of the contests and what one player would do.

This is called utility. In mathematical game theory, utility is comprised of the set of preferences of each player, organized hierarchically in a numerical scale. For that reason, utility has been also said to be preference of payoff function. If the letter *u* represents utility and the letters *a* and *b* indicate the options that will be given numerical values, the utility rule can be formulated as follows: $u\ (a) > u\ (b)$ *if and only if the player prefers a to b.*

Utility is then a matter of degree, allowing the transformation of the fuzzy idea of preference into a quantifiable hierarchy.

The more desirable the preference, the higher the number, for utility is individual preference quantitatively measurable in a numerical progression. Ever since von Neumann and Morgenstern's formulation of the notion of a game, the concept of utility has been used to cover a wide spectrum of preferences: some preferences are better than others; thus they have more utility. Straight numerical expression of a preference is not all that important; what matters is that "for one person one utility is greater than another" (von Neumann and Morgenstern 1944: 16). The consequence is that utility can mean that the players' choices may represent complex and contradictory variants as is the case of the father playing chess with the son; however contradictory it may seem, in this case, the defeat will be a victory for the father. All his efforts must be to lose the game and therefore to obtain his highest utility.

Complexity is not itself a theoretical virtue. Instead, complexity demands reduction and simplification or utility would be an impractical and untenable concept reduced to the role of a crude metaphor. Utility can indicate bewildering complexity, but that is not always the case as one sees when shifting its discussion away from human situations to the world of biology. Perceived or intended sense of personal victory or defeat can be disputable; number of offspring cannot be questioned. Utility is, in this case unambiguous. That is John Maynard Smith's (1982) point. Utility is useful for biology because it can be expressed in very simple terms: the organism wants to replicate itself genetically with greater or less frequency. In the universe of biology, the criterion of strict economic rationality turns out to be population dynamics.

It is, however, quite true that individual preferences may be extremely subjective to the point of being impossible to compare them with ease and objectivity; but that does not invalidate the explanatory power of utility to determine how an individual decision, whether animal or human, could be made. Although not necessarily expressed in numbers, utility is decisively present in the individual player's mind. As long as the computation of preferences is possible, utility is a relevant consideration to define a strategic choice in a game. Conversely, without the definition of the player's utility one cannot evaluate its choice of strategy. Utility is therefore a primitive notion in game theory; it precedes the reaching of an outcome, and consequently its enforcement. From the recognition of individual utilities, the convergence of the interests of the players comes about. Utility reframes and expands the anatomy of the game. It still is an interaction of basically selfish players, whose actions are determined by what the adversary may do, which in turn means that they will follow a plan of action: a strategy. Nonetheless, the plan of action should be drafted according to the payoff preferences, the utility, of each player. Utility determines the actions that a player is willing to perform.

6.3. Adding up to zero

Forgetting for a while the dazzling minutiae of players concretely interacting with one another, it is important to begin commenting on a type of game that, being an irreducible conflict of interest, deserved the privileged attention of John van Neumann: zero-sum games. In fact, one can speculate that von Neumann would have admitted if told that the original model of the zero-sum game should not be poker, like he was always trumpeting, but the game between players of the same sex in dispute for a mate. If zero-sum games are the ones in which one scored point implies necessarily a point lost for the opponent, sexual dispute between males is a mode of zero-sum game. The winner displaces the other competitor. Its offspring is a loss - albeit temporary - for the sexual competitor.

The utility of a player is in direct opposition to the utility of the contender. At least in one way, poker is just like sexual procreation: the chips are on the table; the winner takes all; the losers get nothing. Presume that the poker round has one set of chips on the table; this amount results from the process of matching bets, quantified in whatever number of dollars. The winner takes all the chips that he put on the table plus the ones that were matched as the betting progressed. Losers not only lose their chips; they gain nothing. If one gained chip (or point) is added to one lost chip (or point), the result is zero: (+1) + (-1) = 0. All zero-sum games are of that kind. The matrix below is a graphic depiction of the outcomes of zero-sum games with two players and two results: one player wins a point and the other loses it:

+1, -1	-1, +1
-1, +1	+1, -1

While von Neumann conceded that in any game with multiple players, an n-person game, the players could make alliances, agreements, and communicate, he could not conceive of the incursion of cooperation in zero-sum games. Without considering cooperative acts, the application of the understanding of zero-sum games to social life presents major problems. Restricted to zero-sum interactions, game theory would only deal with bellicose situations. The link between game theory and society's webs of cooperation requires more than a helter-skelter of gains and losses.

Apparently, since in a zero-sum game the player who loses cannot gain anything, zero-sum games are disjunctive and do not foster sociability. That is, however, a false assumption. Situations of extreme conflict like warfare can mature into occasions of mutualism and cooperation. Just think of an armistice, a truce, or the phenomenon of "live and let live" during World War I that Robert Axelrod (1984: 73–87) analyzed so acutely. Enemy troops stuck in trenches and insurmountable frontiers, incapable of moving forward or backward, decided not to shoot one other much to the dismay of their commanders. Then, how could one trace a strict line separating competition from cooperation even between enemies? To demonstrate its worth as an interpretation of social life, game theory has to explain not only competition and zero-sum games, but foremost cooperative interactions and non-zero sum games.

The objectives just raised are in no way minor and indicate that the thoughts around zero-sum games are in potential contradiction with the objective of many aspects of social life, with, for instance, economic activities, whose goal is to create wealth for the individual that, indeed, can be advantageous to others. For example, the shopkeeper who offers specials benefits from the customer's money; and the customer will get goods for a lower price. It is not a win/lose situation for either of them: the shopkeeper and the costumer win in different ways. Biological life is also peppered with activities that are regularly and directly dependent upon cooperation. How is the ordinary social reality of cooperation conciliated with the cutthroat practice of a zero-sum game such as playing poker? In playing poker the winner appropriates what is available over the table. A poker victory is just transference of values, not the creation of anything. Of course, sharing defeats the purpose of zero-sum games. Now the dilemma is if, despite such steadfast objections, one should still have something to learn from zero-sum games. Zero-sum games may be artificial constructions more relevant to parlor games than social interaction, but studying them makes it possible to explain the persistence of deceitful messages and at the same time clarify under what conditions is worthwhile to bluff.

At its inception, game theory dealt only with extreme competition. Von Neumann and Morgenstern's *The Theory of Games and Economic Behavior* (1944) focused its analysis on zero-sum games; and all zero-sum games have an unambiguous non-cooperative and non-altruistic slant. The difficulty in seeing how game theory qualifies to be a social theory is understandable. Game theory seemed to be unashamedly committed to individualism and selfishness.

As each player in the game aims at defeating the adversary, both *A* and *B* chose strategies that maximize their gains and at the same time minimize their losses. That is the solution logically required when selfish players play a zero-sum game. John von Neumann and Morgestern call it the minimax criterion. Such principle is equally valid for equally endowed optimizers. So just like the right hand of an onlooker is reflected as the left one in a mirror, minimax mirrors maximini. Against *A* playing minimax, *B*'s strategy should be maximini. *B* maximizes its gains, minimizing the points that *A* may score. In practical terms, each contender tries to anticipate the adversary's move. The preponderance of the minimax-maximini strategies in zero-sum games is such that its dominance precedes the actual development of the game itself. The success in zero-sum games requests the application of inverted minimax strategies.

In a poker game, *A* has been served a set of cards. *A* has already decided that in the case of a good hand, he would strategically bet 10 dollars. Consider that *A* wins and *B* loses. The catch is that the strategy cannot be repeated: B will easily guess that every time *A* bets 10 dollars, he has a good hand. Applying the maximini solution, *B* passes and does not bet. It is obvious that the same insight is available to *A*. Pure strategies - the ones that establish "always do that" or "never do that"- are frequently ineffectual in zero-sum games. Mixed strategies – the ones that play "this" 60% of the time, for instance, and play "that" the rest of the game – are much better suited to zero-sum games. The adversary has trouble predicting what *A*'s move will be. In zero-sum games, information must be withheld from the adversary; therefore, it is only rational to bluff.

The optimal strategy in poker would be to bluff always with a bad hand, but never to bluff with a slightly better hand (Binmore 1990: 4). The adversary will be confused, incapable of separating a better hand from not so bad cards. Bluffing players should block the comprehension of the adversaries, making their guesses more difficult. The less prediction is allowed to an opponent so much the better. Curiously, von Neumann's theoretical suggestion of never bluffing with a slightly better hand demands a higher standard of bluffing than most players are willing to do. One reason for that is that game theory is normative, not the result of an inductive abstraction of players' actual tendencies to behave at a poker table. Von Neumann's suggestion is better suited to a universe of player where each plays optimally.

Game theory mainly indicates what a player ought to do to solve the problem. Its task is to show what *type* of strategy ought to be developed optimally. The rest is the individual player's responsibility and a product of personal skill. Another important notion is how game theory evaluates strategies, formalizing them in mathematical terms, without noticing that strategies are really embodied in sequence of signs. Messages traded between players are decisive features linking strategies to the outcome of games, either as a victory or a defeat, a point gained or a point lost.

6.4. Pennies for your thoughts

Much less complicated than poker, and a favorite of many introductory books to game theory - perhaps because more illustrative of the peculiarities of zero-sum games - is a game that has been called *matching pennies*.[2] Like poker, matching pennies also presumes two players in direct opposition to each other. It is a game of conflict of interest. The players will simultaneously flip a coin. If coins are the same (head/head, tail/tail), player *A* wins; but if they are mismatched (head/tail, tail/head) player *B* takes all the money from the table, as it ought to be in any zero-sum game. The next matrix is an extension of the one before, and shows that matching pennies is a zero-sum game.

	Head	Tail
Head	+1, -1	-1, +1
Tail	-1, +1	+1, -1

It is perfectly feasible to illustrate zero-sum games without conventionalist overtones. Using the analogy of combinations of matching pennies, imagine that an organism has two potential sets of immunizing passwords composed of two chemical elements. The organism would be closed to the invasion of parasites if it carried a uniform block of chemicals, such as x/x, or y/y. That would be just like head/head, and tail/tail; but a mismatched coupling - x/y and y/x, comparable to head/tail, and tail/head –means that the parasites could open the immunizing lock and invade the organism. Conventions and cultural rules are not at all imperative in the analysis of such games. The emphasis here on matching pennies is merely didactic. The conclusions about matching pennies are equally valid in the case of the immunizing chemical passwords.

Sheer chance always rules the first move of matching pennies. Presume for argument's sake that the mismatch of pennies was the final result from turning the coins simultaneously. It is tail/head and *B* wins. Now it is time for the second round. If *A* is compelled to repeat the strategy, and B is forced to do the same, *B* obviously wins again: however, that is not fair. With freedom of choice in the case of matching pennies or with sexual mutation in the natural world where parasites and immune defense compete with each other, the situation could be quite different. *A* could change its strategy, and then it would place heads over the table, perhaps thinking that *B* would do the same. But suppose that *B* thinks – and rightly so - that *A* will choose head; this choice would be tail and tail comes up again, benefiting *B*. Of course, the outcome could be inverted. The whole purpose of the game is to outguess the adversary, or to diverge materially in the case of immune defense.

A striking peculiarity of matching pennies is that heads and tails are devoid of any value in the game; they are neutral alternatives equivalent to the extreme of making the player indifferent to both. The same happens with chemical defense. So to choose one or the other is not enticing in itself. They remain always just equally probable options. To aggravate this trend of indifference, the players must turn the coins simultaneously, therefore making it extremely difficult to make an educated guess of what the adversary might do. In any case, the players should try to avoid being figured by the adversary, who can try to foresee the future from previous patterns of action. It is not important to try to discover the opponent's intention. Hiding oneself from the adversary is much better (von Neumman and Morgenstern 1944: 144). At this point, the clear inference about the game is that its solution ought to be "play random" ("have random immune defenses"), use a mixed strategy, but so mixed that no one can guess, not even you, the player who is turning the coins. The optimal strategy is to toss a coin to choose the side of the penny to be placed on the table. No one, not even the player flipping the coin to matched would know what will happen.

In tic-tac-toe, the outlook is different. The moves of the players are not simultaneous. One player chooses where to put his mark (an *x* or an *o*) after the other has played. Chance is not operative in tic-tac-toe. The player makes his choices knowing fully and unequivocally what the adversary had done. The choice is grounded upon *complete information*. Being impossible to bluff, the result of tic-tac-toe tends to be a draw. Zero-sum games cannot thrive fully with perfect information. It is no surprise that the game of tic-tac-toe ends frequently with a tedious and predictable tie. Then why play it if to win depends totally on the mistakes of the adversary, much more than on the individual skill of the player? Superficially, the blame for the banality of tic-tac-toe lies solely on the inadequate blend of a zero-sum game with complete information; but that would be a thoughtless conclusion. Chess players also make decisions based on complete information, however with a significant difference: as the game progress, the checkerboard becomes emptier, opening up alternatives. The chess player cannot foresee what direction the game will take: there is no complete list of alternatives available to anyone of the players. Although the list is not infinite, it is quite high: "there are 20 possibilities for the adversary's first move, some 400 after the next, etc." (Sigmund 1995: 161).Tic-tac-toe has nine initial squares; and as the game advances, the nine initial alternatives are gradually reduced. Tic-tac-toe is an easy game at which to guess. Regardless of its being a game of complete information, chess is much more interesting. Chess fascinates people, but not tic-tac-toe.

Zero-sum games must allow for players to deceive the opponent. Self-protection, selfishness, and deception came together. That is why, in considering zero-sum games, Von Neumann and Morgenstern recommended the delivery of messages designed to induce the opponent to act against its own self-interest. The mixed strategy proposed in zero-sum games suggests the necessity of conveying incomplete information to block the adversary's move. The semiotics of deception and incomplete information dominates zero-sum games. How about chess? In this zero-sum game, there is complete information and still the game is not trite. The interest in chess is that, although a game with complete information, the prediction of a chess move is really impossible. Information is the key to prediction. Chess, von Neumann argued coherently, can only be called loosely a game. In a conversation with Jacob Bronowski (1992: 432), von Neumann defined chess as a well-defined form of computation; if one cannot find the correct solution, it does not mean that it does not exist: there is always the correct procedure for that specific position of the pieces on the chessboard. If only the player can make the computation, a solution is there waiting. The player's participation is restricted to complying with the suggestions coming from the rule. In such a case, there is no room for individual maneuvering. Obedience to what the rules recommend is more important.

6.5. Ruling the game

Some zero-sum games – like bridge, poker, or chess - appear to be defined by rules that determine the strategy to be applied during the match. While that may be superficially true in the case of chess, it is a far-fetched dogma to presume that rules cover simultaneously the description and the prescriptions of the game. Conventionalists dismiss the creative role of individual players, fostering the illusion that rules, descriptions, and prescriptions overlap. For that reason, the description of a strategy can create the impression of being a complete product of rules. The inference is that rules precede, allow, and justify the interactions of the game. That is an idea akin to Ferdinand de Saussure's basic conception of sign in *Cours de Lingustique Générale* (1916). The language-structure (*la langue*) precedes the individual speech, just like in a music score one will find instructions to perform the piece, leading methodologically to the conviction that the system of rules must be the privileged object of semiotic studies and to the acceptance of a blind cultural determinism.

The error in this reasoning is that rules are conceived as predetermining the nuances and the purpose of playing a game. If one believes that rules are the prescriptive essence of a game, it is tempting to agree with the anthropological reading that Claude Lévi-Strauss (1962b: 48) makes of Saussurian semiotics in the first chapter of *La Pensée Sauvage*. Lévi-Strauss compares games to rituals. Games and rituals are similar: they are social practices made according to fundamental rules. Without rules, games and rituals could not exist. For him, games distinguish subgroups within the general group, singling out – after the result –the winners and the losers.

Lévi-Strauss insists: rituals are the inverse of games. Rituals unite individuals to the totality of the group. Games divide. Through rituals the group is unified into one. This is as Durkheimean as it can be. However, in conforming to rules the individuals participating in ceremonies display their personal qualities and social worth, either dancing, playing music or simply abiding by other rituals To presume that a ritual is done initially for the benefit of a group is mere idealization, closer to the holistic guiding-hypotheses of the anthropologist than to the actual reality of social actors. A social actor performs a ritual to forge a distinction between him and the other members of the group. Performance harvests prestige and improves the chances of an individual being sexually selected. Not only games, but rituals also divide.

Returning to the previous example, although von Neumann tried to separate chess form other zero-sum games, a chess game dramatizes the individual limits of the player. The loser is incapable of finding the prefect and ideal move to decide the game. The winner was capable of identifying the correct movements and the position of the pieces. What matters is the relative excellence of competing players: a game with mediocre players is just the lukewarm exercise of complying to rules.

More interesting than the recognition that rules are an element of a game is to provide an explanation concerning the persistent frequency of rules in economic and social transactions that otherwise should be guided strictly by individual interests. Why in situations as diverse as trading values in the stock market, buying a product in a shop, or haggling in an ancient Oriental bazaar, is it deemed that the relevant institutions be relatively stable and well-established so that the players do not have the desire or the necessity of violate their norms? (Binnmore 1990: 8). For sure, violation of institutional rules is not an absolute impossibility, but more of a deviant exception than the recurrent habit.

The contention that the stability of institutions has the function of exclusively favoring the solidity, the constancy, or even the identity of the group is indefensible. A group is the result of individual interacting, and it could not precede individual players. The stability of institutions and the permanence of relatively fixed rules are conditions for all the economic agents to perceive that while the individuals may be tempted to try to win at all costs, only the social actors of the finest quality can afford not to yield to the immediate and cheap temptation to fraud the economic system. Most of the actors comply with the rules to avoid being labeled as having low social worth.

In games, another phenomenon deserves note. So that a dispute between players be leveled, and the competition fair, the outstanding, initial disparities between the contenders must be diminished. That is why weight classifies boxers. Fighters will then go to the ring to face contenders of their own category. Something similar happens in economic transactions. Suppose that someone wants to buy a used car at a car dealership. The buyer knows little about cars and their condition. The dealer is not only an economic agent who knows about cars, but also has had the opportunity to examine thoroughly the car that the buyer may want to purchase. The seller has the upper hand.

To offer a warranty for a certain period of time from the purchase date of the vehicle reduces the unevenness of information inherent to this transaction, which otherwise favors the seller. Reducing the risk involved in buying a used car, the dealer offers the warranty to demonstrate that he is correct, honest, and trustworthy. In a situation like this, of voluntary economic exchange, the buyer as well as the seller feels that the trading is individually favorable to both. The seller volunteers to reduce his initial advantage offering a warranty based on the information that he has about the car. In complete opposition to what is frequently done in zero-sum games where complete information must not be disclosed, the asymmetry of information is diminished, and the transaction may flow more easily along cooperative lines. The buyer is encouraged to purchase the vehicle. For that reason, the institution of warranties (or, in another context, the quality offered by brands) is made into a mechanism promoting the equilibrium of conflicting interests of the buyer and the seller. To leave the market free to a point that its only principle is predatory profiteering backfires as self-destructive. The unfettered selling of bad products is against the common interests of buyers and sellers, although the seller will feel the pinch only later. Acting in a predatory manner, the seller will increasingly see the hedges of the used car market shrink to a point that will drive away most potential buyers, except the very wealthy. Then what is to be done with the army of used cars? The institution of a point of equilibrium between buyers and sellers is beneficial to everyone.

Equilibrium reduces market uncertainty. Akerlof (1970) began his analysis on the effects of quality on market uncertainty thinking about used car dealerships, expanding it to insurance companies, then to the contract of labor force among unprivileged groups as well to loaning procedures in communities without great financial resources. For each situation of conflict and asymmetry, institutional mechanisms of equilibrium are created to react against the structural imbalance between players. Points of equilibrium emerge, demonstrating a notable transformation of the minimax criterion, which seemed to von Neumann and Morgenstern to be the exclusive standard of all games. Without suppressing self-centered individuality, cooperation between selfish players evolves.

6.6. In equilibrium

Twenty years before the publication of Akerlof's paper – "A market for lemons" – dealing with the reaching of an equilibrium point in apparent conflicts of interests through the reduction of asymmetric information, the forerunner in the theoretical question of economic equilibrium had been John F. Nash, Jr. His work is nowadays considered the theoretical landmark in the understanding of the complex relationship between competition and cooperation. It is indeed the "basis for almost all work on the refinements of equilibrium concepts" (Mirowski 1992: 115). When he wrote his groundbreaking paper "The bargaining problem" in 1950, Nash (1997: 7–13) had the ambition of providing a drastic revision of von Neumann and Morgenstern's assertive conception of games.

The bargaining problem was particularly relevant to his aim because it is not a lofty abstraction of a conflicting situation metaphorically represented as a poker game, like von Neumann did with economic transactions. Bargaining is one of many daily interactions in which players with conflicting interests do profit from their mutual transactions although none of them may have absolute and maximum gains or even an equal advantage split between them. It is not just the case of buying in sales or getting rid of accumulated seasonal merchandise or even the interaction of someone who wants to buy a used car and the dealer who sells it. It is what occurs in sexual procreation, where females and males have mutual pleasure and, if pregnancy comes, both mates spread their genes to future generations. Males and females do gain something, though not in the same proportion. The examples just mentioned specify that competition and cooperation do not exclude one another. An equilibrium point is the optimal solution for both players despite their constitutive dispute of interests.

To come to this conclusion, Nash initially adopted von Neumann and Morgenstern's presumptions. An economic transaction would be the accomplishment of a strategy, and this strategy was supposed to be evaluated whether or not it was an optimal solution. If von Neumann and Morgenstern's reliance on zero-sum games led just to a complete transference of payoffs, Nash imagined what he called non-zero sum two-person games. A non-zero sum game is not a whole transference of payoffs. The addition between payoffs is not zero, and it reflects the fact that both economic agents profited from the transaction.

Compared to von Neumann and Morgenstern's configuration of zero-sum games, the question of bargaining is not merely the resignation to an outcome that cannot be changed. The bargaining solution is a matter of mutual consent between the players; and consent is possible because of the amount of utility satisfaction for the contenders, not just to one, but also to both players. Nash brought the sparkle of sociability into game theory.

The ingenuity of John Nash's solution started with his particular conviction that an axiomatic approach could be the key to the complex problem of bargaining that had baffled economists for quite a long time. Since the axiomatic method depends on the drafting of general principles from which conclusions will be deduced, Nash's initial step would have to be the drafting of general principles; but such principles should differ from John von Neumann's solution to zero-sum games. Taking into account his studies with David Hilbert, it is easy to understand Von Neumann's consistent adoption of Kantian mathematical formalism. Von Neumann's early years with Hilbert had driven him to impose axiomatic reasoning to several domains of applied mathematics, whether quantum physics, probability theory, game theory or the neuron study of the brain (Mirowski 1992: 126); von Neumann was, however, a strong influence that Nash would have to leave behind.

Although von Neumann's conception of a game situation started from zero-sum conflicts, he was really assessing axiomatic game strategies. The conflict itself would be then left behind, for the overall purpose was simply to evaluate the effectiveness of plans of actions that would be the solutions to games. As Mirowoski (1992) noticed in an excellent historical review of the intention of the collaboration between von Neumann and Morgenstern, just like all his applications of mathematics to everyday problems, Von Neumann had been searching for a modular, strictly mathematical, and rule-oriented game strategy, and that was the meaning of his recommendation of bluffing and mixed strategies to solve the conflict. Starting with von Neumann's assumptions but departing from them, Nash also stipulated that to solve the problem of bargaining, it is necessary to reframe it in axioms that presume the existence of two self-centered players with opportunities to collaborate in more than one way. The issue becomes how this collaboration is reached. That is an important initial supposition because bargaining is a mutual exchange between players quite different from the cutthroat definition of zero-sum games always on the brink of bringing about absolute gain or total loss for the players.

Being a two-person game, the bargaining model postulates the interaction of two rational individuals, equally zealous of their personal interests. They have equal bargaining skills and complete information about the facts and the purpose of the games as well as full knowledge of the tastes and the preferences of the adversary. Here, complete information is not an unreasonable abstraction. With increased time for the transaction, any initial state of incomplete information is gradually reduced. Then why not start from the assumption that the players have - because they can have - complete information? Equally rational players with the same bargaining skills will necessarily engage in an arms race that leads through learning to a state of full and complete knowledge. In that case, bluffing is not even a remote possibility.

A poker game may last for a while, but it is a sequence of brief interactions irreducible to one another. Poker is the eternal beginning of a new round of conflicts. The players can learn about the players at the table, but poker is not an accumulative arms race. Players may try to guess the actual strategy from the knowledge of previous actions; it is, however, just a precarious guess. Deceitful signs and the fluctuation of whims in players are undeniable possibilities in real life settings with actual interactors, but, for simplicity's sake, they had to be expurgated from Nash's model.

With such assumptions, Nash's "The bargaining problem" was able to disregard the discrepancies of actual players and focus its analysis entirely on strategy; otherwise, all the previous insistence on consent and satisfaction could have ended in fuzzy qualitative assertion. Nash argued that from the notion of utility, it was viable to grasp the intricacies of bargaining.

In bargaining, a player must be able to anticipate with certainty what alternatives he can get. According to Nash, suppose that a car is about to be given to a player. He knows that it can be a Buick of a Cadillac. The player does not know yet what car he will receive, but he can scale his preferences and assess the possibility of his satisfaction in the case of each gift. Such scaling is numerical: his full preference is for one vehicle, but the other falls somewhere in between no satisfaction (zero) or full satisfaction (one). The situation is expressed in this way: $0=<p=<1$. The point p may not be what the player ideally wants, but it is not negligible at all: p is better than zero. Mathematically, one of the brilliant results of Nash's paper is his application of Kakutani's fixed-point theorem to economic analysis. Using the fixed-point theorem, Nash establishes an area of anticipation where one player's maximum point of advantage would meet another's maximal point of advantages. That is where the equilibrium point lies.

The players' means of judging what they can get is transitive: *A* is better than *B*; *B* is better than *A*; thus *A* is better than *C*. If this reasoning occurs in both players' minds, although they may differ in their utility, there must be a satisfactory utility function of both players. If not, there is no mutual consensual point of agreement. Bargaining is successful if agreement and mutual cooperation is reached, and conversely fails if that is not the case. Formally bargaining can happen and by extension, cooperation if a fundamental condition is met. In the particular set of alternatives for each player, there is a point C of conflict of interest, which is also where bargaining as a solution can take place as long as the utility is greater than the advantages of the conflict: U1>=C1 and U2>=C2 (Harsanyi 1989). If that is the case, the agreement can take place.

As said before, to provide an alternative to von Neumann and Morgenstern's conception of zero-sum game, Nash could not be satisfied with the mere transference of wealth from player to player. Each player, considering synchronically his own utility in relation to the utility of his adversaries as well as potential points of conflict, will realize that there is a point that justifies his participation in the bargaining game; this is the point of equilibrium: a point from which none of the players set on maximizing their gains will have a reason to depart. That is the first shadow of the celebrated Nash equilibrium. Theoretically, the Nash equilibrium was a total reversal of the previously dominant conception of zero-sum games.

After Nash, the minimax criterion that had been seen as the cornerstone of game theory was redefined as a case of Nash equilibrium. Von Neumann and Morgenstern's previous recommendation of using mixed strategies to bluff other players was in the Nash equilibrium case quite counterproductive. Messages could be trustworthy; and semiotic interaction with complete information did not lead at all to inevitable and entire losses. In a bargaining situation, both players expect to gain something with transparent messages displaying complete information; this is not a mutual altruistic surrender. *A* is trying to get as much profit from the bargain as *B*, and *B* is doing the same; the bargaining problem is solved if both players get their maximal gains. If *A* and *B* are rational, equally intelligent and capable agents, all bluffing between them is neutralized. Bargaining cannot occur in an ambience without cooperation and trust.

How does this fare in a highly competitive setting? In *Non-cooperative Games* – his dissertation of 1951 – Nash (in Kuhn and Nassar 2002: 51–84) inquired into the possibility of games where it would be impossible to establish alliances or to communicate with the other players to make any agreement that would benefit the players in the game. Paradoxically, Nash demonstrated mathematically that in such a highly antagonistic scenario as long as the games are finite, there is always at least one point of equilibrium; and in any finite game there are symmetrical points of equilibrium. Equilibria underline zero-sum games.

6.7. Cutting and choosing the slices of a magical pizza

What one deducts from Nash's early papers on conflict and non-cooperative games is that a zero-sum game has at least one equilibrium point. Equilibrium is an optimal solution for both players; and in a state of equilibrium none of the antagonists has any reason to depart from the solution. The bargaining solution is optimal because it fulfills the criterion of mutual satisfaction.

A simple and popular illustration of all this is the case of two players trying to gain as much advantage as possible in the cutting a pizza. The absolutely selfish ideal is to take the whole pizza away; but that is not possible, because the opponent also wants a slice of pizza. What is the reasonable expectation of the two players? What maximum gain can each player expect when there is conflict interest? As in all game situations, from matching pennies to poker to sex, the reply to this question is frequency-dependent, thus contingent upon the mechanics of interaction within a set of players, indeed on what the opponents do.

Suppose also that the conflict of interest is represented structurally in the division of functions of each player. Player *A* cuts the pizza. Player *B* chooses in sequence the slice. The cutter thinks on what to do. Cutting the pan in unequal slices is a weak strategy, one easily countered by the opponent. The chooser takes the larger slice and leaves A, the cutter, with a solution that is certainly less than optimal. For *A*, the ideal solution is to cut the pizza as close as possible to the middle. So *B* – the one to choose – will have his gain as minimized as possible. It is typical minimax situation. But the interesting point is that the minimax does not lead to any gain; or – what is more significant – both gain half of the pizza. When minimax and maximini add up to zero, the point of equilibrium is just like cutting the pizza in half: a magically fair solution of equity without the imposition of an external enforcer. The example displays equality without the abandonment of self-interest. The minimal advantage for one player is the maximal advantage for the other player.

If player *B* were the one to cut, the solution would be the same. To cut in half is the most robust strategy, the one that no opponent can invade. In the case of a divided pizza, the point of equilibrium is in the division as close as possible to the middle, not because of a lofty sense of justice and altruism – always a risky hypothesis – but because it attends to the selfish interests of *A* and *B*.

A game has Nash equilibrium when the solution of each player is optimal for all the players despite their conflicting interests. To deviate from the strategy is a worse solution than to stick to the previous state. It is an optimal strategy for every one of the players in the game; and as such it is firmly hedged against any other alternative. The 1:1 male/female ratio is the Nash equilibrium for sexual beings. John Maynard Smith (1982) called this phenomenon an evolutionary stable strategy (EES). It is from a situation of equilibrium that social rules can evolve. At this point reciprocity is socially legitimate. Without an equilibrium point it is impossible to have lasting reciprocal rules; and staunch enforcement becomes mandatory. Interestingly, the minimax criterion – albeit its extreme competitiveness – works as a center from which comes an equilibrium point that does not contradict the interests of both players. Dismissing the selfish interest of the contenders in the social game is implausible, because selflessness is an idealization.

In the illustration of the sharing of a pizza with two players with excluding functions – one cutting, and the other choosing – the solution of the problem is dividing the pan as closely as possible to the middle. With three players, one cutting, and the others choosing, the solution is to cut the pizza in three equal slices, and so on progressively; therefore, with a set of n-persons the pizza will shrink to nothing; but that is not the only issue when the number of players increase even if the pizza increases proportionally to the addition of players. In large clusters, players have little or noting to gain and to loose; and no equilibrium points are established among the players. Anomie becomes a threat. Small groups foster sociability.

The unity and the force of the group over players is directly proportional to the number of players in the game. The Arcadian anthropological dream of a stable and permanent culture derives from the limits of the interpreted social groups, if not from the fact that the anthropologist had to rely too heavily on a small number of informers. When the number of players in the game increases, then more utilities must be taken into account. Therefore, the stability of mutual equilibrium points shown in Nash's bargaining model with two players simply splinters off. There is always a player who does not like these pizza toppings, and would rather have something different; there is someone who is on a diet, and others who simply hate pizza, and so on. The growth of the numbers of players multiplies points of equilibrium. In such a situation, the games can only be played if the players keep on communicating with one another incessantly, establishing multiple and unstable points of social balance. It is not that signs are individual performances from prior collective rules: society is based upon semiotic exchange. The existence of a unified and unifying equilibrium is nothing but an idealized formulation that only exists in the simplicity of theoretical models. Endless acts of communication and ceaseless negotiation substitute for the demand of a centralized enforcer.

Originally published in 1953 in *Econometrica*, Nash's (see Kuhn and Nassar 2002: 99–114) "Two-person cooperative games" develops the ideas of "The bargaining problem." Nash's "Two-person cooperative games" reveals that a plan of rational bargaining presumes the discussion, or a form of communication, of individual viewpoints. After that, threats are conveyed to the other players of the game. In a latter stage of the process, the threats are carried out. That assertion is a step ahead toward grasping the role of communication in games: threats are the decisive semiotic foundation of a bargaining game. Whether merely believed, or actually executed, threats are the sign movements that wrap up the conflict. Accepting a threat as unworthy of being challenged amounts to the maintenance of social equilibrium. Although Nash kept himself within the confines of a two-person game, it is quite reasonable to imagine societies as nets of equilibrium points made and unmade by veiled or explicit threats that weave the multiplicity of individually unstable viewpoints.

Is it possible to reach a unified collective order from the sum of individually conflicting interests? How can one decide what incomparable utilities must be excluded without a control center that responds for the convergence of individual interests of the players? Plato affirms unequivocally that the center is not only necessary, but it does exist in the player who knows more and can coordinate social diversity: the philosopher-king. More recently one can find comfort in the Durkheimean notion of society as a moralizing and cohesive technique of social control.

However, if both Plato's and Durkheim's utopias are unacceptable, it is impossible to come to any notion of coordination top-down without paying the heavy price of social control. Social coordination is just the establishment of a tyranny. In this case, the situation is quite similar to what Kenneth Arrow (1983: 24) claimed in his theorem establishing the impossibility of having a social decision that expresses a common point of collective interest. When individuals are the starting point, it is impossible to determine a center in which one can find the inclusive expression of collective interests. That does not happen in voting, and certainly not in markets. While it is true that individual preferences can be computed, in the end what this computation does is just to project the ghostly shadow of a unifying collective preference that does not exist at all in the social fabric. All social situations are under the sway of unstable equilibrium. Social equilibrium is only reached through incessant semiotic interactions in cultural arenas.

Chapter 7
Players

Before sinking into a paralyzing mental illness, John F. Nash, Jr., drafted with a handful of papers a whole program of mathematical and economic research. Without Nash's active participation in the long, subsequent debates of other theorists, his contribution to the discussion on economic equilibrium could have slowly faded into oblivion, but this prospect was turned around: his name and shadow are now fully projected over the whole field of forces that came to constitute contemporary game theory.

Just like John von Neumann and Oskar Morgenstern's *The Theory of Games and Economic Behavior* (1944), Nash's program gives absolute privilege to the evaluation of optimal strategies. Indeed, since Nash's bargaining model presumes the equal sharing of complete information, nothing is hidden from the players; no whims or personal quirks interferes with the process; no mistakes move the interested parties away from the foreordained goal of getting to a deal that is the optimal solution for both of them. Not only is Nash's conception of the bargaining problem extremely elegant, but it is also neat, and, perhaps, too tidy.

In a peculiar manner, the mathematics of the bargaining model sanitizes the reality of satisfaction in mutual transactions. Nash's "The Bargaining Problem" is artfully unconcerned with how complete information about the players' utilities is obtained. The set of preferences of each player is public domain prior to the closing of the transaction. Nash's idealized teleology suppresses the sway and the uncertainty of haggling. Bargaining is the confluence of interests. In Nash's model, the players are always willing to cooperate. The most obvious consequence of Nash' axiomatic definitions of bargaining is that the semiotic exchange between players was reduced to the utmost insignificance; and not surprisingly, for Nash's bargaining model proposes only a methodology to evaluate abstract strategies. Everything else is out of the picture.

Von Neumann's axiomatic search for a rule-governed solution to economic conflicts finds full expression in Nash's idea of equilibrium. Zero-sum games and their reliance on the minimax-maximini criterion are made into a case of Nash equilibrium. Any conflict has at least one point of balance from which the players can afford not to depart. Cooperation pays off because the losses do not outweigh the rewards. The best is the least worse. After Nash, although in apparent contradiction with the inevitable self-interest of players, it is possible to see how cooperation could have developed across the natural world.

7.1. The storm blast came

Despite the differences in their solutions to economic games, von Neumann and Nash rely on a common idea of rationality that starts with players facing a zero-sum situation. Each player is equally competent. Unflagging self-interest dictates their actions.

Although von Neumann claims that it is rational not to trust the opponent, and therefore the players should adopt the mixed strategy of bluffing to win the game, Nash' s conception of equilibrium takes the optimal solution to another level. An optimal bargain implies the satisfaction of both players. How can one forget that the opponent is equally endowed and capable of counter-reaction? Yet, putting the preferences of each player on a gradual scale and comparing them reveals what they would have to do to achieve mutual satisfaction.

A player is risking his own self-interest if he goes for an enormous reward and dismisses what the opponent might do. Strategies that pursue absolute rewards without considering the opponents' actions and gains could well be invaded. For Nash, rationality and mutual satisfaction overlap entirely.

Actual bargaining is not as straightforward and as immediate as Nash makes it look. Nash's shining theory has one blemish: it assumes that a rational decision depends upon the psychological quality of the players. That is in contradiction with the strictly rational requirement of game theory. Accepting equilibrium in an economic transaction in which one player gains something, but the opponents' reward is much higher presumes more than rationality. It requires resignation.

Resignation may be perfectly acceptable under specific conditions. Consumers accept a merchant's margin of profit as long as the prices offered are not exceedingly high in absolute terms or in relation to the other sellers. If a merchant's products are too pricey, the consumer searches for better values. Consumers take care of their own self-interest, maximizing their gains while minimizing the sellers' margins of profit. That is primarily rational. It has nothing do with psychological traits. Consumers also agree that it is acceptable to purchase a product bought for a certain wholesale price and sold in the market for another value if the difference between the price bought and the price sold is not perceived as outrageous, that is, if the consumer does not get the sucker's payoff. More often than not, the conflict of interest between buyers and sellers is not really a problem. Everyone knows that the seller aims at selling as high as possible to consumers that want to buy goods as low as possible. Old and robust minimax prevails.

Disputing this assumption – after Nash had published his paper on bargaining – Merrill M. Flood (1958) asked himself what defined the fair and mutually satisfactory price of a good. His initial guess was that if the difference between price bought and price sold is split between a buyer and a seller, the outcome is rational and mutual satisfaction.

Imagine this example: a car is usually sold at a dealership for $ 20,000. A person who wants to sell his car bought the vehicle for $ 17,000. The difference between the same product bought and sold is $ 3,000. The equilibrium point is an equal split between the product sold and the product bought. It would amount to $1,500. The fair and acceptable price of the object in the same condition should be as close as possible to $ 18,500 ($17,000 + $1,500). That seems to put an end to establishing the fair price of reselling the product, especially if one considers that the individual seller does not have the usual costs of operating a dealership. That is not the case at all. Any price between the hypothetical top value of $20,000 and the lower range of $ 18,501 is a potential equilibrium point. Like so many economic games and unlike the previous example of cutting and choosing a slice of pizza, this game has more than one equilibrium point. So, what is the criterion to fix the satisfactory price for both buyer and seller?

In Flood's analysis of experimental games, the two situations were contemplated: one of them with one equilibrium point, and the other with a wider region of choices. The first illustration is the selling of a Buick. Before the players find an equilibrium point, a fair price for seller and buyer, or even any scale of values that can allow for the success of their transaction, they have to be willing to reach a mutual agreement. They come to an agreement because they want to agree. The situation is akin to the circularity of conventionalist arguments: conventions need prior conventions to reach a convention.

Coalitions between players can only be established after the players exchange messages, whether demonstrating good will, vowing allegiances, advertising motives, or delivering threats. The "splitting-the-difference" principle has to be displayed and discussed between the players. When a coalition is impossible, the finding of mutual satisfaction is an unattainable goal. To show that a theory of haggling is a mandatory condition for bargaining, Flood (1958: 9) presents the case of buying and selling an Oldsmobile. The two economic agents do not agree on one fixed priced. They haggle within a range of prices. The deal is eventually closed, but the buyer ends up with an Oldsmobile with a price slightly larger than he was expecting to pay. The seller holds his ground, thus getting a better price than splitting the difference.

The real fixing of a satisfactory price with multiple equilibrium points is dependent upon the action of players. This is quite disturbing for mathematicians, who always like to have one solution – the correct one – for the problem. For them, an array of alternative results is usually seen as the sign of deficient reasoning;[1] but not only that: consider the simultaneous possibility of the seller wanting to get as close as possible to the $ 20,000 top value, and the buyer trying to lower it as much as possible. Presume also that they are stubborn players. The haggling goes on, backward and forward. In the process, the players gather more than necessary information about each other. They maintain their initial bargaining positions coming nowhere close to each other; and then the buyer concludes that it is useless. The seller is perceived as unreasonable. The buyer walks out of the deal. Both economic agents lose, for we presume that buyer and seller have the same wish: one needs the car; the other must sell the product.

Why does the deal fall through? Is it just because of the buyer's insistence on the highest price? Not really. The price is high but it was within the equilibrium range. Or is it because the economic agents have gathered information about the other player? Other factors that escape economic rationality play a role in the example. Perhaps the players' mutual dislike erects a barrier between them. The seller could have described the pristine state of the car, but the buyer sees him as a greedy and pushy salesman. If that were so, no matter how reasonable the price is, the deal will not happen. Optimal and rational bargaining can be easily subjected to irrational forces that get in the way of the economic transaction. Such reasoning is completely out of von Neumann's and Nash's considerations. They only cared for strategies. For them, strategies function as rules capable of guiding the individual behavior of economic agents. The assumption is that the players merely go by the directive of rational decisions.

Another of Flood's interesting experimental games puts a dent on the convergence of von Neumann's minimax and Nash's bargaining equilibrium. Flood offered to share a $150 cash prize between two of his secretaries at RAND (Poundstone 1990: 102). He wanted to see if the equilibrium mechanism of splitting the difference created a sense of fairness and justice. He proposed to Secretary 1 the prize of $100 plus $25 if she convinced Secretary 2 to accept the remaining 25 dollars. The game is one of complete information: both would know what was going on.

The transaction itself is not unfair at all. Secretary 1 would certainly leave the deal with more money, but she would have the burden of convincing Secretary 2. Secretary 2 should be attracted to the deal; she would get $25 for almost nothing just accepting the proposition. If not, both secretaries would get zero dollars. For each, it was better to accept the deal than just to refuse it. In a typical equilibrium situation, Secretary 1 and Secretary 2 gains, although – like in any bargaining situation with different utilities – the rewards are uneven. To be uneven does not mean unfairness.

The result of the bargaining between the secretaries is astonishing. They decide not to accept the offer; they would rather take the loss, but preferably, the secretaries would split the $150 even, thus getting $75. It is a kind of splitting, for sure, although not along the original proposition of Flood. The situation is absurdly irrational. The secretaries risked getting nothing. Moreover, Secretary 1 accepts a loss of $50. That is amazing and unreasonable. Not only Secretary 1 was willing to accept the loss of $50, but also by rejecting Flood's offer, Secretary 2 loses an easy gain. How could she doubt the apparently evident advantage going from zero to $25? Refusing to cooperate, holding ground, Secretary 2 manages to transfer $50 from Secretary 1 to herself. This story is the root of a crucial problem in game theory: the celebrated Prisoner's Dilemma.

7.2. A ghastly crew of uncooperative players

Albert W. Tucker, a former teacher of Nash in Princeton, told the little fiction of two imprisoned crooks that we now know as the Prisoner's Dilemma. The tale was an elaborate version of Flood's noncooperative pair of secretaries.

Originally presented to a group of psychology graduate students of Stanford University in May 1950, with minor adaptations, the story is this: the police charge two detainees with a legal infraction. The prisoners are kept in isolation. The district attorney makes them an offer. The prisoners realize that if one confesses, and the other does not, the one who does not provide the evidence to the district attorney will be freed, while the other will get the whole penalty. The one who confesses will get ten years in prison; the one who keeps his mouth shut gets no punishment at all. However, if both confess, each will get five years in jail, splitting the punishment; also, if both cooperate with one another, the two will go clear.

The situation of the prisoners is an obvious conflict of interest. The district attorney's propositions sets one prisoner against the other. They have two alternative strategies: either they cooperate with each other, or they simply betray the comrade in crime, doing what game theorists called defect. Defection is providing the evidence so that the district attorney can put another felon in jail. Unless they manage to stick to the rule of silence, the district attorney will get what he wants, and at least one of them will mutate into a jailbird. As in the previous postulates of game theory, the players are equally selfish and endowed with the same capacity of rational discernment. If *C* represents cooperate and *D* indicates defect, the possible outcomes of the game are:

1. Both cooperate with one another: *C*/*C*.
2. Prisoner 1 decides not to cooperate (he defects), while Prisoner 2 thinks that it is best to do the opposite: *D*/*C*.
3. Prisoner 1 behaves as Prisoner 2 in the previous outcome; thus, they cooperate; Prisoner 2 defects: *C*/*D*.
4. Both prisoners defect: *D*/*D*.

If the prisoners would have to consider those alternatives, the optimally rational solution for the rogues would be the first outcome. They would cooperate with each other and soon hit the streets; but the scale of penalties poses a mutual problem. The district attorney's proposition is a tempting offer to defect. The temptation to defect – value a – is higher than the rewards for cooperation – value b. Value c is the sucker's payoff. Put each value near the other and see the scale of payoffs in the Prisoner's Dilemma: $a>b>c$. The temptation to defect (a) has to be at least higher than the addition of the rewards for cooperation (b) and the sucker's payoff (c): $a>b+c$; and they surely are: $10>5+0$. That proportion of payoffs definitely tips the scale of preferences in the direction of defecting. Note that without this scale of values, the prisoners would face no dilemma whatsoever: they would not be really tempted to defect.

Here comes the puzzle: to cooperate with each other is certainly better than not to cooperate. While not confessing is excellent, the favorable outcome of the alternative of choosing to cooperate – as it should be in a game – is wholly dependent upon the other prisoner's willingness to behave in the same way. There is no other option to save oneself: the prisoner must hedge information from the district attorney; but on what grounds can the prisoner be assured that his partner will do the same? For, as it must be remembered, the partners have no means of communicating with one another, and more important, it is a single deal. The instantaneous decisions are made in the dark. The objective of the progressive scale of the payoff ($a>b>c$: 10>5>0 years of sentencing) is to test the strength of the prisoners. In a situation like this, is it rational to resist the temptation to defect?

The two prisoners clearly realize that one strategy dominates the outcome of the game; and that strategy is to defect. Although an initial cooperative leap is mandatory to an excellent personal outcome, whoever cooperates is liable to get the worse deal if the other defects. Unilateral defection is a much better strategy than one-way cooperation. The only rational strategic choice to protect oneself against the brunt of the worse deal is to defect anyway. It is quite true that the player who defects gets a bad deal – the sharing of the sentence – but that is sill better than getting the whole punishment: $a>c$; $a>b$: 10>0; 10>5.

Oddly enough, the most reasonable strategy for the selfish players is not the ideal one. Without the punishment scale, the prisoners ought to cooperate mutually. Why do they not do it? The prisoners are rational fools (Sen 1977). They rationally abandon their best chance, the mutual chance to cooperate (*C/C*) and be free. That may look crazy, and yet there are reasons to their madness: just compare the strategy of triggering cooperation with the consequences of defecting. It is all so very well to bear in mind that cooperation means freedom for the prisoners, but in fact, that is not the only outcome. Prisoner 1 cooperates and Prisoner 2 defects: the cooperator gets the highest punishment. If Prisoner 1 defects, while the other confesses, and gets the whole punishment: not bad for number 1. Nonetheless, if both defect, they get a bad deal that is still better than running the risk of cooperating. Inevitably, whoever plays defect gets the chance of going free, while the other may rot in jail. Immersed in the task of assessing possibilities, the prisoners realize that defection can be more promising than cooperation. As it is indicated in the Nash equilibrium, the best can be become no more than the least worse.

Considering that the alternative to be prescribed in a game is the one that cannot be defeated, what is the optimal plan of action in the story of the Prisoner's Dilemma? Phrasing it differently: What is the alternative to which there would be no incentive to deviate from the decision? Paradoxically, it is not the ideally best. Indeed, the most rational strategy is the one that bars the achievement of the ideal strategy. Players with equal mental capacity defect mutually and share a penalty when they could have gone free if only they trusted each other.

Certainly more exciting than Flood's narrative of grudging secretaries, the fable with the prisoners' dilemma attracted a lot of attention because of the important problem it poses. In real life, one can see countless examples of individuals who have a selfish nature and still move away from zero-sum games to non-zero situations. Cooperation is made from the crooked timber of selfishness. It may seem a hopeless contradiction, but living organisms are fundamentally selfish, and nonetheless cooperation comes to life from competition.[2] Even among brutish animals, societies are not a rare fact. While the evidence of widespread cooperation is not difficult to see in many instances in the natural world, the problem remains: How do zero-sum games mingle and relate to nonzero modes of cooperation? Or are cooperation instances merely irrational and therefore arbitrarily occasional accidents in life?

The surprising conclusion of the Prisoner's Dilemma is that rational beings are led to behave irrationally. Amartya Sen (1997: 6) put it precisely: "the interdependence between different people's welfare may make the pursuit of individual interests produce inferior results for all, *in terms of those very interests*."[3] That is the astonishing feature of the Prisoner's Dilemma: irrationality is grounded upon reasonable decisions. The irony is that self-centered players are forced to act against their own interests. How can a pair of rational players with the same logical tools, hypothesis, and principles, applying the Nash equilibrium, get to an outcome that is a rather bad payoff? The central hypothesis of Nash – the ideal bargaining point is one where none of the players has any reason to depart – is torn apart. The theoretical promise of the Nash equilibrium is the reaching of the best possible outcome; but in the end, the players are caught in an untenable situation, incapable of optimizing their pay-off in the game.

The lesson seems to be that rational players make irrational decisions. Although they are free to decide on the best strategy, the players are entangled in a strong web that displays their rational frailty and impotence. One of the main attractions of Tucker's story is its humbling effect. A kind of Prisoner's Dilemma is at the core of Michael Corleone's dilemma in Coppola's *The Godfather*. Michael may have rated highly his utility in leaving the seedy underworld of his Mafia family, but self-preservation overruled his intentions. He knew that his enemies could break the truce, so he attacked them. That is truly a dreadful choice for someone who vowed to walk out on organized crime. Michael Corleone had no real choice; he was sucked into a whirlpool in which he never wanted to be: "Just when I thought I was out, they pulled me back in."

The Prisoner's Dilemma is a cautionary tale about the difficulty and the limits of achieving cooperation with self-centered actors. What is the solution to the dilemma? The initial reaction to this question is to see the prisoners' story as a hopeless dead end. The conditions of the game are rigged. Barred from getting to an agreement that would be the solution to the problem, the felons have to try to save their own skin; but even if they could have come to an agreement before their arrest, that would not be enough. One prisoner could promise not to snitch on the other in the case of being caught, and yet the pledge could be simply broken. However, if one prisoner had power over the other, such as having the possibility of killing the comrade's son as retaliation, the one with such power could perhaps risk to trust that the agreement would be kept; but to be a mutual leap into the cooperative move, power would have to be spread evenly. Otherwise, the one with power would be tempted to betray the companion. Strange as it may be, trust is far harder to obtain than raw power. Trust is frequently plagued with uncertainty and often invaded by mistrust.

Consider a recurrent reaction when Tucker's story of the Prisoner's Dilemma is presented to a class for the first time. It is common to hear students saying, "I would not necessarily snitch; but that would depend on who was arrested with me." The discussion persists if the instructor asks how the student knows that it is possible to trust someone absolutely. Some reply, "I know some people that I trust: I trust my brother." It is productive at this point to remind the student that the point of the investigation is to assess strategies not people; but fine, take the case of the brother: Can brothers be always trusted? How about Cain and Abel? How about the ones who snitch on brothers to collect rewards? It would not be unusual to hear from the back of the class: "I would be betrayed in a heartbeat." Now, return to the issue and ask, Why do you trust your brother? "I do so because he is a stand-up guy." That type of reply establishes that the reason for trusting someone is that this person stood the test of time. Trust then emerges as an effect of the players' recurrent interaction. Time is the condition for the evolution of cooperation among selfish players.

7.3. Serving time

The prisoners in Tucker's tale faced a dilemma not just because their utilities are tilted toward the temptation to defect. Conditions under which the interaction unfolds play a major role in the outcome of mutual defection. The game is a single deal. The players must make a decision that amounts possibly to a leap into a dark abyss. How to trust the other players if they cannot communicate with one another? A loose narrative thread in the tale of Tucker's dilemma is why such rogues did not have a pledge of mutual protection, a contract larger than the individuals and strong enough to bind them to a more favorable outcome. Honor among thieves is widespread. Yet, the rogues could be rookies or first-time losers.

The conditions of interaction across time and through semiotic acts of communication are quite reasonable and should be considered as central elements of the game. If the prisoners were haggling, they would interact much more than in the case of instantaneous decisions. They would also have to communicate recurrently. The period of the game ought to be extended far beyond the immediate instant. Prisoner's Dilemma can be played repeatedly, and that is a regular feature of life. Players meet each other repeatedly. Rarely are strict and absolute single deals prevalent in life; and when they occur, they are more often than not irrelevant and soon to be forgotten.

To consider such factors, Robert Axelrod (1984) devised a computer tournament to solve the laborious enigma of the Prisoner's Dilemma. It was tournament of strategies transformed into computer programs set against one another. The competition followed the round-robin mode, which permitted the recurrence of meetings between the strategies. Not only that, the strategies played against each other more than once as well as against itself, thus presenting a complete alternative to the one-shot deal typical of the classical Prisoner's Dilemma. Like almost all game theorists, Axelrod chose to evaluate strategies: players make mistakes and act inconsistently, but strategies do not do that. Axelrod's tournament respected the traditional payoff preferences of the Prisoner's Dilemma: temptation to defect (a) scored higher points than cooperation (b), and cooperation was more valued than the sucker's payoff (c): $a>b>c$. In Axelrod's payoff scale, $a=5$; $b=3$; $c=0$. The temptation to defect is higher than the addition of cooperation with the sucker's payoff: $5>3+0$.

The tournament indirectly tested the general conventionalist assumptions of a Hobbesian view of social interaction. As early as 1651 in

The Leviathan, Thomas Hobbes argued that if the foundation of social interaction is individual interest, the outcome of social life has to be more than solitary: poor, nasty, brutish, and short. The solution to such a curse is to distance society from the natural state. To exist socially, groups of individuals must create or accept social conventions and power structures that indicate a strong central enforcer capable of mediating and curbing the outright conflicts among players. Is an enforcer totally necessary for the emergence of cooperation?

A central feature of Axelrod's game is that the players ought to have memory; otherwise, the very idea of extending the period is senseless. That is fair and acceptable for in the Prisoner's Dilemma with a single deal, time – indeed the instant – is an important element in the players' decision. Without the memory of the previous rounds, the past is excised from the game. With an undefined future and the players devoid of a sense of past, the game is be hopelessly stuck in a eerie eternal present. Since players are endowed with memory, they can abstract the opponent's strategy from previous interactions.

Having established the conditions of the game, Axelrod then invited 15 scholars to submit strategies to the tournament. As a benchmark for the lowest-scoring plan of action, Axelrod introduced the randomized strategy, which amounted to no strategy at all. No plan of action could score lower than Random. In the tournament of 225 games – 15 strategies facing 15 strategies – the minimal score would be zero (not even Random's absence of a plan would score that), while the maximum number of points would be 15,000. To get to the highest score, one must multiply 200 by the 5 points given to value *a* (temptation to defect) and then multiply this result by the number of strategies playing against each other, by 15 plans of action. No player is anticipated to score the minimal (zero) or the maximal (15,000) number of possible points. The expected number of scored points should be realistically somewhere near 600. The strategies would score points game by game. The points would be recorded, and then the scores would be added up to determine which of the submitted strategies would be the winning plan of action.

The winner in Axelrod's first tournament was Anatol Rapoport's submission, a strategy called Tit-for-tat (T-F-T). T-F-T's plan of action was deceptively simple: its first move was always to cooperate, and then do whatever the opponent had done. T-F-T's strategy could be reinterpreted in terms that are more didactic:

1. T-F-T was a mixed strategy. If it were a pure strategy, it would entail always cooperate (ALLC) or always defect (ALLD).
2. T-F-T's specification of its second move – always do what the opponent has done – meant a short-term memory. It remembered just the last round, but it forgot all previous actions of the opponent, regardless of how many times these actions were repeated in the past.
3. In an anthropomorphic vocabulary, T-F-T was forgiving. It had some niceness to it (indeed, forgiveness is a nice trait). T-F-T was a strategy without great resentment. It could bear some grudge; otherwise, it would not remember the previous move of the opponent. Nonetheless, T-F-T did not go overboard; it said to the opponent: although you have done this to me many times before, I will forget it for this round; but if you repeat it, I will strike back.
4. That indicates the extent of T-F-T's retaliatory nature. When the opponent is not nice, T-F-T would not be nice at all. It paid the opponent in the same currency, but not in excess. T-F-T retaliates in moderation.
5. T-F-T was semiotically very simple. According to what Axelrod declared in *The Evolution of Cooperation* (1984), many submissions were consistently complicated sets of instructions. Tit-for-tat's simplicity stood out and thus presented some semiotic advantages: the opponent could easily get it; it communicated well and unambiguously to the other players.

7.4. Unto others

The best way to understand T-F-T triumph is to see its match against other strategies. To keep the explanation as simple as possible, it will be presented here an abstract of Axelrod's tournament, setting typical strategies against one another. With a short description limited to four moves, it is perfectly feasible to assess T-F-T's performance in relation to other strategies.

1. Let us put Tit-for-tat against a nasty and pure strategy, such as always defect: T-F-T x ALLD. The results are *C*/*D*; *D*/*D*; *D*/*D*; and finally *D*/*D*. Altogether, considering the scale of points in the game,
2. T-F-T does not fare well against a nasty strategy such as always defect. Any strategy beginning with defection defeats T-F-T.
3. Now consider an always defect strategy against a strategy similar to itself, a variation of always defect, for simplicity: ALLD x ALLD. The outcomes would be *D*/*D*; *D*/*D*; *D*/*D*; *D*/*D*. The game is a draw. If the

two strategies are mainly but not always committed to defection, the one with a greater number of defections moves wins the match. From the comparison of these results with the previous match, a strategy beginning with defection and developing into cooperation could not do well against ALLD. In any round of *C* against *D*, whoever plays *C* loses.

4. In the case of T-F-T against ALLC, the results would be: *C*/*C*; *C*/*C*; *C*/*C*; *C*/*C*. The points that T-F-T lost in the match against ALLD are now compensated. Encounters between cooperative players are blissfully sucessful.
5. When always defect faces always cooperate, ALLD x ALLC, what we see is mayhem: *D*/*C*; *D*/*C*; *D*/*C*; *D*/*C*. ALLD takes an easy walk, invading ALLC without mercy. The advantages that ALLC gained cooperating with T-F-T and other modes of *C* are completely lost. Reinvigorated, ALLD returned to the game scene more robust then ever. An important point must be made here: to offer the other cheek all the time is a social disaster. It strengthens players with nasty strategies. Absolute submission, total conformity, unconditional cooperation, and all modes of passivity, indeed, any form of extreme sacrifice, is socially harmful: unqualified cooperation incapable of retaliating gives room to strategies uncommitted to niceness. At best, ALLC is shamefully selfish. The one who turns the other cheek may win the Kingdom of Heaven but do at the price of putting the community on the brink of an invasion. Then why some individuals like to praise the virtues of submission and sacrificial obedience? Their goal is to create a cluster of suckers that would be their easy prey. Although not in a pure mode, for no pure strategy is always effective all the time, lack of cooperation, non-conformity, and disobedience are valid social reactions against the hypocritical preachers of obedience and submission. Submission and obedience are always an excellent advice just for the others: the message is *do what I say, not what I do.* Why do suckers themselves keep on spreading the gospel of submission and conformity? To them the bribe is rather tempting: you may be a sucker for a lifespan as brief as a handkerchief, but you will get everlasting paradise. No other special could ever beat that offer.
6. Suppose finally that T-F-T played with a strategy that is not pure but does not begin with cooperation. It will start loosing, but it may win the end. The outcome would depend on the development of the game. To T-F-T, such sets of strategies yield random results: either victories or defeats.

It is not difficult to guess that in the whole tournament T-F-T could score relatively well. T-F-T performs awfully among defectors, but it does much better than the strategy ALLC. With cooperators, most especially ALLC, T-F-T does exceedingly fine. Defectors against defectors will experience trouble, but cooperators playing with cooperators thrive. With any cooperator – including versions of itself such as Tit-for-two-tats (T-F-2-T) – T-F-T manages to do to its best. T-F-T is a more effective strategy than ALLC: T-F-T has a defense mechanism to combat exploitation. However, thrown in a universe with a greater number of nasty defectors, then T-F-T is in a fix. Cooperators ought to play preferably with cooperators. Groups of cooperators with a retaliatory move can barricade ALLC and assail the defectors. It is in their own selfish interest to protect cooperators. Even if some cooperators – the absolute ones – fare worse than a mixed strategy such as T-F-T in their interaction with defectors, the balance of strategies in the population of players must be kept. Imbalances and the sole dominance of one strategy – such as a population without retaliators – tip the scales in the direction of defectors: defectors harm each other ruthlessly, thus fostering the emergence of a population with pure cooperators that is easily invaded.

In a peculiar way, the lesson to be inferred appears to reinforce conventionalist images of social interaction although not along strictly Hobbesian lines. In a community of cooperative individuals, cooperation flourishes. Cooperation breeds cooperation in the same way that conventions bred conventions. Is that a profound assertion of a trite statement?

The circularity of social beings needing social beings to be social did not seem to bother Axelrod. When he considered the results of the first tournament, he was intrigued by the triumph of T-F-T. He then decided to expand the game with a greater number of players, notwithstanding an important variation in the rules: it was not made explicit at what number of moves the game would end. It was just like life, the time of death is unknown.

At the end of the second tournament, T-F-T won again, and Axelrod (1984: 44) noticed that of the 15 strategies that performed best, all but one were nice strategies. A nice strategy is a program of action bent on cooperating with other players. The tempting inference from this is that niceness is a desirable social trait. What is baffling is how the redundancy of sociability and niceness went unnoticed. To be nice is to be social; and social traits are considered nice. Circularity comes again to stiffle what could be learned from Axelrod's tournament. Nice strategies are reliable: they foster trust among other players, and this contributes to the creation of an environment that is collectively stable.

Although nice, a cooperating strategy should not be naïve. Without the capacity to retaliate, nice strategies are easy prey to opportunists and defectors. However, an important trait of T-F-T is that it should not start with an uncooperative move. Nice strategies – the ones that begin cooperating, like T-F-T – look ahead to the future. They are never stuck in the past. Nice strategies run the risk of being nice, and this risk is an assurance of its niceness. Generosity complements the niceness of T-F-T. The message of T-F-T is disarmingly evident: its set of instructions is simple enough to be understood by almost everyone.

Because it contributed to the creation of a lasting environment of cooperators, T-F-T has been hastily qualified as an evolutionary stable strategy. It was robust; it fulfilled the requirement of any Nash equilibrium: it gave no incentive for other strategies to abandon cooperation. When two variations of T-F-T play against each other, the unilateral defection is immediately punished: that is a deterrent to avoid defection from a collectively stable strategy. After T-F-T emerges in the environment, just like sex, it becomes a hedge against invaders. Coupled with other nice strategies, T-F-T creates a world of collaborators. That was the rosy interpretation for the triumph of T-F-T in Axelrod's tournament.

7.5. The tit-for-tat blues

At close range, however, T-F-T displays some obvious weaknesses. If the player knows at what round of interactions the game ends, T-F-T has to assume that the opponent, however nice in the past, will be tempted to defect at the last moment. Therefore, the rule that defines its plan of action – "do what the other did in the previous move" – is ineffective and should change. To survive, T-F-T must not be T-F-T anymore.

In an odd way, if T-F-T were thus altered, all the previous moves, not just the last one, would have to change following the same recursive logic. As an example, suppose *A* plays *B*, both knowing that at the fifth round the game would be over. *A* is T-F-T, and *B* is a cooperator, but programmed to change its strategy at the last round of any game. Of course, the player *A* following T-F-T cannot know that; it only goes back to the previous moves. Let us see the first four rounds and their outcomes: *C*/*C*; *C*/*C*; *C*/*C*; *C*/*C*.

Now, if *A* follows T-F-T's rule, it must play *C* (cooperate); but *B* moves to *D* (defect), wining the game.

Imagine that player *A* (T-F-T) remembers this bad experience and bears in mind that *D* is a strong possibility at the end of the game. Then, it makes no sense to play *C*; not only at the last round, but also in any round before the last: player *B* is not a trustworthy cooperator. The recursive moves goes from the last, back to the first. Again, does it make sense to begin playing *C*? It is senseless to start playing *C*. The first move has to be defect; otherwise, *A* would loose; if *A* begins cooperating it will suffer the same disadvantages that it had with ALLD. Two lessons exist here: first, *A* (T-F-T) must be resigned to lose or it must become another strategy; if not, it will pay the usual price of losing when facing any defector. The idea that T-F-T creates a world of cooperators is no solace. If predominantly cooperative, this cluster of nice players can be invaded by defectors, thus tilting the frequency to a world of defectors. So, if successful, T-F-T becomes its own gravedigger.

One could counter-argue: this would not happen, for life is a game of undefined end. Is that convincing? Any player can behave as though the game has ended and start to defect. Even if the reaction is defect also, the sudden and unilateral defector has the advantage of having decided first, and T-F-T, like any cooperator, just lags behind. T-F-T is vulnerable to this strategy. To survive, it has to be something else but T-F-T. It has to move beyond its initial cooperative trait. T-F-T must be more than merely reactive: it has to be proactive. All such considerations lead to a conclusion rather different from Axelrod's and Hamilton's (1981): it is always best to start with defection and then eventually cooperate if the opponent is a cooperator. That means the adoption of a strategy quite the inverse of T-F-T: do not cooperate at first; begin defecting. It would be as Binmore (1994) contends: Tat-for-tit or Suspicious Tit-for-tat (ST-F-T), in the terminology of Boyd and Loberbaum (1987).

Another problem with T-F-T is the probability of noise in the interaction. Noise – technically defined as anything that distorts the intended content of a sign – can create an unbreakable cycle of mutual recrimination. Returning to the situation of *A* (T-F-T) playing *B* for five rounds. First round: *A* plays *C*, *B* plays *D*, but it is a mistake: what should be a precise signal is really noise. Second round: following the T-F-T's plan of action, *A* plays *D*, and *B* can have two alternatives: it may play *D* or *C*. If *B* plays *C* when *A* plays *D*, then it is obvious that *B* gets the message – altogether wrong – that *A* is not willing to cooperate. The cycle can go on for quite a long time, if not forever, with both players imprisoned in their recurrent strategies of defection.

If the time of repeated interactions is large enough, T-F-T can be invaded eventually. Axelrod's analysis presumed that a collectively stable strategy could be deemed an evolutionary stable strategy. An evolutionary stable strategy, however, must be more than collectively stable. Evolutionary stable strategies should be measured by their fitness against invaders. In a population of meek, ALLC doves, being a dove is collectively stable but not necessarily fit enough to resist the action of a single ALLD, nasty, hawkish strategy. Boyd and Loberbaum (1987: 58) defined the difference properly: to be evolutionarily stable, the common strategy must have higher expected fitness than any rare invading strategy. T-F-T does not meet that criterion.

In terms of assessing the fate of evolutionary stable strategies, it is quite acceptable to presume that a mutation of T-F-T can develop into what Boyd and Loberbaum (1987) called Suspicious Tit-for-tat. ST-F-T is clearly not like T-F-T. It is its inversion. T-F-T's first move is always never defect; ST-F-T begins defecting before starting to cooperate. ST-F-T is robust: it takes advantage of T-F-T weak point and starts the game wining. While T-F-T can react to the first unilateral defection of ST-F-T, in the end every move of T-F-T is simply one step behind. ST-F-T meets the criterion that defines an evolutionarily stable strategy: it resists invasion from rare and nasty strategies. The first move is self-protective and functions as a hedge against unpredictable players. Indeed, not all players can be predicted in their first move. T-F-T's initial cooperative step is ineffective and gullible beyond words.

Had T-F-T mutated to a more generous type, such as Tit-for-two-tats (T-F-2-T), the overall picture would have changed. ST-F-T would have had a more cooperative relationship with T-F-2-T. ST-F-T would begin defecting, but that would not send T-F-2-T into a retaliatory frenzy. T-F-2-T would have held its ground, and the cooperation could develop in the rounds to come. The two strategies can then cooperate with one another, which was not possible with T-F-T. The variety of strategies neatly corresponds to the variety that sex introduces in biological life. With T-F-T, ST-F-T does better than with T-F-2-T. Their coexistence is evolutionarily stable.

T-F-T may carry in its structure the means to break the potentially endless sequence of recriminations. One must always remember: T-F-T is forgiving. The crux of the issue is: How forgiving? T-F-T cannot be forgiving all the time; otherwise, it is ALLC. Should T-F-T forgive one mistake and then, retaliate in the next? Is T-F-2-T the solution? It is not enough to forgive just single mistakes, but to forgive recurrently. In this case, the question is: How occasionally, and in what proportion? Could it be not two, but three strikes, and you are out? How generous should be Generous-tit-for-tat (GT-F-T)?

In an article dealing with the problem of noise in the repeated interactions of the Prisoner's Dilemma (Wu and Axelrod 1995), Axelrod examined some of the developments that could fine-tune T-F-T's strategy. Generosity is thus seen as a trait that is programmed for ten percent of the time, allowing for mistakes and not departing too much from what T-F-T usually does. Contrition is another attribute, this time just to deal not with the opponent's mistake, but with T-F-T's own errors. Contrite Tit-for-tat (CT-F-T) does not respond to the other player's defection if it is not a unilateral defection. CT-F-T recognizes its own errors and does not retaliate if the other defects, thus not getting into endless and mutual recrimination. If the opponent cooperates, CT-F-T reposes in contentment; but if the antagonist defects, the strategy reacts to the provocation. Generous Tit-for-tat (GT-F-T) and Contrite Tit-for-tats (C-F-2-t) have their own limitations: Generosity counters errors of the other player, while contrition deals with its own errors. GT-F-T and CT-F-T are complementary strategies, which means that neither is a fully effective strategy in dealing with noise. Moreover, even if GT-F-T were combined with CT-F-T, why is GT-F-T forgiving 10% of the time? What defines this percentage is not clear at all.

Binmore (1994: 197) observed acutely that if Prisoner's Dilemma is played indefinitely, the game has an infinite number of equilibrium points. They can be: defection against defection, Tit-for-tat against Tit-for-tat, absolutely unforgiving T-F-T (also known as Grim) against Grim. All such pairings are Nash equilibria. The presence of countless Nash points of equilibrium makes it virtually impossible to evaluate the most appropriate strategy for the game.

7.6. Someone's gotta give

Defection can take more than one form and meaning. It can be a protective measure to hedge the player against the invasion of another player that may defect as well. In a game where the dominant strategy is defection, the possibility that this may be the opponent's choice is too real and too imminent to dismiss. Despite the contention and the recognition that cooperation may evolve among reciprocating players, no pure strategy is ever stable in a population playing interminable Prisoner's Dilemma. Defection remains a strong possibility in such a situation. The achievement of T-F-T in Axelrod's championship of strategies underscored this. T-F-T had as one of its strategic choices the possibility of retaliation; the mixture of cooperation and defection set T-F-T apart from gullible programs such as ALLC. What is in question in the previous critique of T-F-T is never its mixed nature, but whether defection should come as the first move. Symptomatically, the same dually stable phenomenon manifests itself in the realm of signs: deceptive representations are never wiped away although cooperation and trust may be temporarily dominant in any semiotic universe.

However, another possibility must be considered. The player may defect systematically. This point is made in the recollection that T-F-T is not just forgiving, but it can also get into a whirlwind of grudges. Having a grudge is definitely not an effective resistance to unrelenting defectors; but what a barricade of grudges may create is outright, persistent, solved, or unresolved conflicts with more or less stability. The defector, in this case, is not reacting, but positively acting to create a niche of its own. Unpleasant and uncomfortable as the insight may be, cooperating is never dominant in a population of selfish individuals because of its frailty and tendency to be invaded. The dual stability of strategies is another indication that evolution has found it hard to reach a decision about adopting or not a strategy in a game of ruthlessly selfish individuals without considering more than just the plan of actions available to the players. The variety that sex introduces in life generates all kind of strategies, and the possibility of having to face nasty strategies is a deterrent from the dominance of purely cooperative plans of action in games. Plans of action succeed or fail in relation to actual players. Ultimately, players – not strategies – are the decisive factors in a conflict of interest.

While the matches in Axelrod's tournament were always made in the dark, with players unaware of what the opponent may do, the initial move of T-F-T is a disarming self-revelation. The assumption is that the succession of moves – the "long shadow of the future," to use Axelrod's famous expression – should allow the opponent to learn who T-F-T is. That again would depend upon the individual capacity of the players. The regular focus of game theory on strategic decisions presumes that the opponent of a player following a T-F-T plan of action could understand T-F-T's moves, for both players have the same logical capabilities. The possibility of T-F-T's victory goes back to the players' individual competency in defining the outcome of the game. In any case, the hope is that if the opponent knows against whom it is playing, it will certainly cooperate because cooperation selfishly yields higher points; but the unambiguous motive of T-F-T behavior is the conditions under which the game develops.

The game in Axelrod's competition was extensive over time, but it was no more than a series of individual matches. The rounds began repeatedly, and it was thus quite similar to von Neumann's poker games. Just like poker, the result of an isolated match did not echo directly into the future; it was either computed in a separate roll or carried along as information processed within the individual players. The accumulation of matches required the active storage of the results of the previous moves – at least the last one – in the player's mind and memory. The players could only come to understand the opponent's traits in reference to its previous moves. No information was available before the first interaction. In this case, the threat of defection was never fully dispelled. Suspicious Tit-for-tat (ST-F-T) is always more effective in such situations. Nothing changed: ST-F-T is still the optimal option in Prisoner's Dilemma, if played outside of the frame of Axelrod's game. Moreover, even with the resource of a deep and extensive memory, the knowledge of a player's past performance gives no exact assurance that the opponent will cooperate. At best, the past points to a more or less drawn sketch of a future trend. Nothing is certain. Axelrod's brandishing of the term "long shadow of the future" does not alter the fact that the future is almost always veiled in complete darknes and that none of the players can fully fathom what will come next. When the players trade signs and what is achieved is imperfect information, T-F-T performs below the range of programs of action that start defecting.

The future's dark veil is so inscrutable that the players will tend to go by what happened in the past. With all that in mind, players in repetitive cooperative interactions may occasionally share an intentionally cooperative mood, but repetition of past moves is not a guarantee of mutual trust. An interesting hypothesis is that T-F-T's decision of being cooperative may be grounded in the role of the risk involved in the first cooperative move. T-F-T takes the wilder chance, and only someone truly set on cooperating would behave like that. T-F-T's action adheres to the principle that costly signs are more truthful than cheap ones.

Axelrod recognized that T-F-T was trying to build its reputation as a cooperator. In *The Evolution of Cooperation* (1984), Axelrod argued that Tit-for-tat may be a somewhat advantageous strategy, but the best way to construct a reputation in Prisoner's Dilemma should be by displaying unequivocally the player's willingness and capacity to defect. With the game establishing that both players must move simultaneously, the initial step would have to be necessarily in the dark. The logic that decided the optimal first move came from an assessment of the dominant strategy in the game, which has been noted previously, is always the imperative to defect.

Therefore, Axelrod argued, the most effective reputation, the one that once built could define conflicts is not the one of cooperator but that of being a bully. Bullies run over cooperators and intimidate other defectors. Not differently from the determination of what is truthful, to be a bully is not cheap. For a bully, "until your reputation is well established, you are likely to have to get into a lot of unrewarding contests of will" (Axelrod 1984: 153). Axelrod's formulation of how the rank of a bully is reached is simply unconvincing. His mistake rests on the preconceived idea that the status of players depends exclusively on past interactions. Bullying – indeed any intimidation – can be far more than just a matter of the past, or even a projection into the future. Bullying and intimidation are reached according to instantaneous and ever-present semiotic displays that advertise the player's personal qualities.

7.7. It is not yellow; it is Chicken

It may well be that the game of Prisoner's Dilemma is just a re-enactment of another game, a game that also sets players against players in a complete conflict of interest, a zero-sum game, where hawks (defectors) have an inherent advantage over doves (cooperators). This fundamental game has defection, risk, and hawkish-ness as the modes of the dominant strategy of the game. The game does not obey the dramatic script of snitching prisoners; its scenario is quite another and reverberates over countless other games so that one could suspect that it is *the game of life.*

Prisoner's Dilemma is one of the polymorphous variations of a game called Chicken. Sometime ago, the game of Chicken was a constant part of the plots of many movies about the rebellious American youth of the 50s. The most famous of them is *Rebel without a Cause,* starring James Dean, although the present canonical description of Chicken is not exactly the one shown in the film.

In the standard description of Chicken, two adolescents drive cars going forward to each other. If the cars do not swerve, the collision could be fatal. The game is decided when one of the drivers gets scared, and drives away, avoiding the crash. The one who swerves loses and is dubbed "Chicken." In the game of Chicken, to be irrational is the foundation of the game's rationality. More evidently than Prisoner's Dilemma, but just like it, Chicken is a game that builds prestige and creates reputation.

The intense drama of Chicken has condemned it to be written off as an oddity, a freakish practice or as an exaggeration. In *Common Sense and the Nuclear War*, published in 1959, a time when Chicken was in fashion among the young, Bertrand Russell used the image of the game of Chicken to describe the insanity of the threat of nuclear conflicts during the Cold War. Nuclear brinksmanship was a derivation of what he qualified as a favorite sport of "some youthful degenerates" (Russell 1968: 30). The sport of bored and irresponsible boys was similar to the nuclear stalemate produced by common nuclear threat: in both games, total destruction became imminent. Worse still, what separated a rebellious youth sport from the political confrontation between nations was that the outcome of nuclear brinksmanship was far more destructive than Chicken. The inherent risk of Chicken was multiplied many times over in the case of a nuclear attack: hundreds of millions of people would be erased from the planet. Russell's view of Chicken was hopelessly negative. It was synonym with reckless and irresponsible child-like behavior. It is, thus, no surprise that – with few exceptions[4] – the game of Chicken has not been taken seriously.

Thomas C. Schelling (1960) holds a more complex view of brinksmanship than Russell does. For Schelling, brinksmanship is quite different from the prank of adolescents but, indeed, part of a serious strategy of conflict that involves the mechanism of tacit bargaining. The players in tacit bargaining decide what potential outcome to accept as a result of assessing the opponent's behavior: they act in accordance to the created expectation. Brinksmanship is not the use of force; it is "the expectation of potential force" (Schelling 1960: 5). The strength of brinksmanship is directly proportional to the risk involved in the game situation. The objective is to get so close to a disastrous outcome that the enemy is lead to avoid conflict, for it is in his own interest. The uncertainty of the outcome makes brinksmanship effective and persuasive. "In a sense, it may make sense to keep the enemy guessing as long as we are not trying to keep him guessing about our own situation" (Schelling 1960: 201). Brinksmanship controls the enemy through the semiotic display of apparent lack of control.

Leave behind the jeering crowds of rowdy young men, and see what kind of game Chicken is. It is an absolute zero-sum game. If one does not die or is severely maimed, the player who loses has his social identity disfigured. One player's reputation is put on the table against another's. If the player is renamed "chicken," he is subjugated and humiliated. He loses face, undergoing a kind of social death; if he insists on going beyond the point of collision without fear or trembling, he may not be chicken, but he may be killed. With Chicken, von Neumann's fascination for zero-sum contests as models for other human practices returned to the game scene under a ferocious guise. Chicken is no parlor game. Chicken is an issue of life and death. Whoever plays Chicken is playing an extended and more serious Russian roulette. It is the danger of dying magnified unto others.

In Chicken, relentless and fearless escalation is a condition for getting a victory or losing everything. The scale of extremes in Chicken is rather simple: if one does not win, one dies or one loses face. To survive the crash is no consolation. The player is harmed, but he is not the winner. Although the crash may be feared, it may be better (for the young man's parents) than death; collision does not establish the outcome of the game, and even less desirable is to avoid death, for it means losing the contest.

The paradox in Chicken is that avoiding death is a sign of weakness. The great players of Chicken are willing to pay the highest personal payoff. Chicken presents a gnarled scale of utility: life itself is at stake; and to stick to life means getting the loser's payoff. The player is simultaneously the winner and the payoff. For those who play Chicken, life should not be prized in excess. To win is to be willing to give your life away, for the simple reason that if the player chooses to live and avoids a collision with players stronger than him, he is the loser. The rule of the game states it clearly: "the loser is defined as that individual who swerves first in order to avoid a head-on collision" (Riechert and Hammerstein 1984: 384). The winner is willing to pay the ultimate cost, and that is one reason for the fascination of Chicken. Another fascinating side to Chicken is that the winner is not a dumb or a traditional bully. Contrary to Axelrod's suggestion, to be the bully is not the best reputation for players of Prisoner's Dilemma. Bullies may fall into disrepute if one opponent unmasks their frailty: bullies cannot ever recover their once lost reputations, but players with a reputation of winning Chicken are individuals of the highest rank, players who survived through an activity of the highest risk. Of course, winning Chicken builds reputation not because of physical intimidation, but because the player has the resolve and the inner strength to run a huge – indeed the highest – risk. Nothing is more important than winning, not even life itself.

Chicken is a game with two Nash equilibria, but because of its extremely radical payoff, the game cannot be identical to Prisoner's Dilemma.

Even a long sentence in jail is nothing like dying. Nonetheless, in Chicken, the Nash points of equilibrium are:

a) Player *A* swerves when player *B* does not.
b) Player *B* swerves when player *A* does not.

From the viewpoint of the payoffs, the potential situations of the game are the following: (i) Player *A* drives on, *B* swerves; (ii) *A* swerves, *B* drives on; (iii) *A* swerves, *B* swerves; and (iv) *A* drives head on, *B* also does the same. For player *A*, the scales of payoff must be: [(i)>(iii)>(ii)>(iv)]. For *B*, the scale of payoffs is: [(ii)>(iii)>(i)>(iv)]. The symmetry of the players reasoning with the same capacity and with the same goal of winning should lead to the common outcome of payoff (iv) – obviously the most costly result – and thus paradoxically to the total extermination of the players and moreover of all those who ever played or will play Chicken. But the fact is that under different cloaks, Chicken is played frequently. If the extermination of both players were common and truly inevitable, evolution would have weeded out such extreme form of competition, for Chicken is the game in which the absolute price may be paid.

The game of Chicken may allow – theoretically, at least – for the symmetry of players, but in reality it is reasonable to expect in the contest that players should not follow exactly the same strategic plan. The most desirable outcomes in Chicken are the ones in which the players adopt different courses of actions. The same action does not define the game: same actions leads to a tie (both swerve) or to death (both collide). In Chicken, a tie is not desirable at all. The outcome of the game must be either victory or defeat. Notice, again, that in Chicken, the two Nash equilibria are the ones in which one player does the opposite of the other. Only when this is achieved, the game ends in a victory. While Prisoner's Dilemma is a game of submission (two men in jail), Chicken is a game whose outcome is the establishment of a relationship of power and dishonor (the winner calls the loser "chicken"). Unlike Prisoner's Dilemma, Chicken does more than submit one or both players to a central enforcer: it adds shame to capitulation.

Direct or disguised acts of potency shape the outcome of any game of Chicken. While in Prisoner's Dilemma one player has no power over the other, and when the game is over, the players remain without force or power over the opponent, Chicken demonstrates who is hierarchically superior and displays clout. Player *A* must dissuade *B* from going headlong in his direction and *B* should do the same. Consider one more time how this situation is different from Prisoner's Dilemma: Prisoner's Dilemma is a subset of Chicken. In Prisoner's Dilemma, the first move is always made in a state of ignorance: both players move in complete simultaneity.

Compare then the reasons in each game why ST-F-T defects initially, and T-F-T does not. In Prisoner's Dilemma, ST-F-T decides its action based on the dominance of the strategy "defection" in the game; but T-F-T's intention is to display its willingness to cooperate. Plunged into complete ignorance, whatever ST-F-T and T-F-T will face is impossible to know. In Chicken, however, the situation is quite another: the players must be asymmetrical in their fierce resolve and determination to pay the highest cost and run the utmost risk. Not necessarily before the players move, but certainly during the unfolding of the game of Chicken, until the last moment before the crash, player *A* must induce fear in player *B*, thus forcing *B* to back off from his initial intention to drive head on and not swerve. For his part, *B* intends to achieve the same.

One player may suspect that the other is bluffing and will not go all the way to collision, but it is never clear where the point of swerving is; and mainly it is impossible to be sure if the other player is so strong that he has no breaking point at all. For some rare and insane players, death may be better than shame. The success of *A* (or of *B*, for that matter) depends not only on his choice of driving straight, but also of intimidating *B* to retract to a less than optimally desired strategy. The solution of Chicken is not in the driving, but in the signs that are sent – prior to the outcome – to the opponent. Whoever wins delivers information whose role is to force the other players to do differently than what he will do. The winner is the player who can do that.

7.8. Signs of asymmetry and asymmetric players

The earliest record of a Chicken situation in the Western tradition is found in Book 23 of Homer's *The Iliad.* After Achilles slaughtered Hector, he dragged the corpse around the walls of Troy, publicly defiling the Trojan hero. It was the ultimate offense for a Greek: the Greeks were extremely demanding about funeral rites. Having attained revenge, Achilles must bury his friend Patroclus. Achilles was happy, but nonetheless plagued by the feeling that his promise of vengeance was only partially fulfilled. For Achilles, the death of Hector was not enough, so he quenched more of his rage cutting the throats of a dozen of young Trojans, whose bodies were thrown into the hungry flames of the funeral pyre that was devouring Patroclus' body. Achilles then announced a series of athletic matches to celebrate the glory of his deceased brother-in-arms. Achilles seemed transformed as if revenge had brought him back from the isolation of his rage and grief for the death of Patroclus to the jovial community of his fellow-fighters. He became a courteous referee, a fair and generous promoter of the games; yet, after the funeral games, he fell back into the same fury, dragging Hector's body three times around Patroclus' burial site. That scene faithfully represents the moves of T-F-T's sequence of cooperation and retaliation, niceness and nastiness. Then to express the honor due to Patroclus, Achilles offered many rich prizes to whoever would win the funeral games. The highest – the most coveted prize – was reserved for the chariot race. The winner would have "a woman to lead away, flawless, skilled in the crafts" (*Iliad* 23, 301). Blatant sexuality underscored the competition.

The racers took their positions drawn by the lot. At this point, as it had to be, equality was absolute. Rules established the conditions of the dispute, and all they did was coordinate the players: as is always the case, rules could never determine the winner. The winner – the most skilled player – would be known later. The race started when, after the signal, the racers altogether lashed their horses. The cars shot on, jolting, whirling dust; the drivers yelled at their pair of horses. Eumelus began first in the race in front of Diomedes, Antilochus, and Menelaus.[5]

Amidst noise and cheers, the race proceeded with the active and treacherous interference of the gods Apollo and Athena. Eumelus's car finally fell apart: it was one of Athena's tricks to counter Apollo's meddling in the race. Eumelus careened off the track, tumbling to the ground. After that, Homer's narrative focused on the dispute between Antilochus and the red-haired Menelaus.

The dispute of Antilochus and Menelaus was, in Homer's narrative, a dramatic game of Chicken. Although the race was an n-person game with four players, *The Iliad* focused on the interplay of Antilochus and Menelaus, making it a two-person zero-sum game, with the racers aiming at the first prize. Antilochus approached Menelaus from behind, whipping fiercely his horses, threatening and taunting them. Antilochus hollered at the animals, that they would not only be punished if they falter, but on the top of that they would be ashamed not just of losing this race, but also of being defeated by Blazes, Menelaus's mare, of all things, a female. The whole race – process and prize – exhaled sex. As Antilochus gained on

Menelaus, getting ready to overcome him, he laid his plan of action to himself and to his horses: "faster, full gallop, I'll find a way, I've just the skill to slip past when the tracks narrow" (*Iliad* 23, 463–465).

Antilochus and Menelaus were ready to play Chicken. They were approaching the place where the track became a sunken rut. Being a rut, it is obvious that Menelaus, Antilochus, and all the other competitors had complete knowledge of this point of the track. Indeed Menelaus screamed, reminding Antilochus that the track would widen soon. It was a classical case of common knowledge in game theory: Antiolochus knew that Menelaus knew that Antiolochus knew; but Antilochus did not refrain his horses, he lashed them even more, in fact completely ignoring what Menelaus screamed at him, behaving "like a man stone deaf" (*Iliad* 23, 480), thus holding firmly his ground. Then, Menelaus was forced to chicken out, frightened of a collision; he began to be left behind, with the consolation of mere cursing: "you drive like a maniac, you're treacherous"; but Antilochus just flogged the horses with greater ferocity, taking the lead that Menelaus was never going to get back. Antilochus ended up in second place to Diomedes, who had been the first all along, since Athena's intervention.

With common knowledge of the tracks, relatively equal determination, balanced proficiency, and skill in manning their chariots, Antilochus and Menelaus should have crashed, but they did not collide because in the dispute more than conditions and strategies were at work. As the players are asymmetrical, games of Chicken do not necessarily end in mutual destruction and collision, although mayhem is always a possibility. The solution of the game is totally an affair of the asymmetry between the players, even when strategies and conditions are symmetrical.

Grudgingly, when the race was over, but defensively, Menelaus acknowledged their differences as players. Antilochus' success was not just the result of choosing a strategy. Something else triggered his plan of action, and it was an effect of his personal quality as a player. When the prizes started to be collected, Menaelaus challenged Antilochus to pledge an oath that he never blocked the rut intentionally, in other words, that Antilochus' strategy was not deliberate foul play. Antilochus did not make the oath, and – with a tinge of irony – offered the compensation of giving his earned prize to Menelaus. Menelaus, however, could never accept it. It would be like contracting a debt, and he would be humiliated even more; thus, he excused Antilochus for having behaved that way, saying that it was his turn to yield to the younger racer, despite his mounting anger: "you were never wild or reckless in the past. It is only youth that got the better of your discretion" (*Iliad* 23, 671).

The stark asymmetry was completely recognized; what can be more contrasting and asymmetrical than the opposition of youth and maturity? Antilochus won because his strategy of playing Chicken was at one with the impetuosity of being young. Menelaus felt that Antilochus would not deviate from his choice of strategy. For Menelaus, the signs delivered by Antilochus during the race were reliable and instrumental in his decision to avoid the clash of cars. Menelaus quit the race as he inferred the daring and the briskness of his opponent, and to reach that decision, his starting points were the indices flooding the competition.

Indices are sprawled over the natural world, and they are not fakes. Indices are not arbitrary and spring from a direct relation of cause and effect. No other representation is as close to its object as an index. Indices are the natural prose of physical events, spelling the truth-content of signs. Facing indexical signs, the interpreters – the players in the game – cannot reasonably doubt the reality of their existence. In another context, forgetting for a moment the Homeric race, a player sees the dry boughs of a tree projecting tortured shadows over a crumbling wall. The player does not turn to check if the tree is really at his back. The paws of a stray dog leave daisy-prints on the drying cement. Shadows, footprints, the creaking of wheels, the lashing of whips, the mane of horses as unfurled sails, and the dust of the tracks are all indices. They are the tracks of physical happenings: they may change and be unstable if their model is altered, but the relationship between the representation and the model is one of close reversibility. Change the model, the representation changes, but the closeness of model and representation is undeniable. It is from such cues (Hasson 1994) that the players develop their strategies, whose goal is always to alter the behavior of the opponent. Indices can function as sponsors of performances. The physical nature of an index is an assurance of its reliability. Indices spring from "an existential relation" (Peirce 1931: 2.283). The physical and the social worlds are covered with indices.

Indices compete with other indices. For an index to be efficient, it must outdo the surrounding representations. All indices have a cost and tend to escalate. The capacity of semiotic inflation is vital for any material form to become a sign. After meeting this efficiency cost, a sign can be a cognitive tool; and cognition is much more than the mere interpretation of representations stemming from shared conventions. If cognition is the knowledge of an object that suffers the constraints of external reality, it can never be a self-centered, independent universe exclusively anchored in arbitrary and codified signs that mediate one another.

If one cannot conceive cognition as the complex and intricate interaction of indices, material procedures, logical inferences, rules, fallible interpretations, themselves made precarious by their inevitable discrepancy with the external world they intend to represent, then one is left with a mere interplay of representations mediating each other. This is precisely the conception of knowledge that carries conventionalist semiotics into a tautological sorry-go-round, thus producing the effects of infinite regress, of perpetual self-reference, so typical of solipsistic semiotics. If rigid conventionalist theories were right, the purpose of examining signs to acquire cognition would be defeated. Cognition from convention establishes a more or less wide range of possibilities, but does not determine which one will occur. Yet, from indices pointing to the determination of Antilochus' resolve, Menelaus inferred an exact judgment about his rival. The collected indices (the fastness of the horses, their relentless whipping, the screams of the driver, the fact that the younger player did not turn to respond to Menelaus warnings of danger indicated indifference to risk and danger. The signs constituted a set of data (*P*) from which a conditional inference can be made: "If *P*, then *Q*." Although the collision had not yet occurred, Menelaus's inference acquired a status akin to a real event. If all those indices (like any group of indices) are real, then the collision should occur. The passage from the "can" to the foreboding sense of "ought" is strong enough to make Menelaus back off. At this moment, the game is solved; the winner has chickened the other player out of the game.

7.9. Types, tokens, and inflated signs

From the viewpoint of time, it is possible to understand why Prisoner's Dilemma is a case of Chicken. Prisoner's Dilemma played once has to end rationally in mutual defection; only if repeated can the game yield room for the relative success of T-F-T. Yet, that is a puzzling conclusion: T-F-T cannot win a single match, but when its points are accumulated, T-F-T fares better than the other strategies. How could such a constant loser end up victorious? Taking into account its victories and defeats, T-F-T had an unsatisfactory performance. If each single match in the tournament were progressively fatal – as it can be so frequently in biological struggles – T-F-T would have most likely perished in the tournament. Accumulation and repetition of encounters is certainly a feature of the evolutionary process, but another central Darwinian insight was left out of Axelrod's tournament. To bet on the "long shadow of the future" requires first having a degree of certainty that an encounter may not be fatal. Dead players have no future ahead of them.

The future or any draft of evolutionary processes is drawn from the success of the players in the instantaneous present. Life is a succession of last minutes, and minutes are a succession of last seconds. The regression goes to the flickering of a fiat over which life hangs on. Of all games, only Chicken can capture this ordeal. The game must go on until the smallest fraction of time before the crash. Up to the last moment, the players can turn the game in their favor, and they ought to try that, for the payoff is either survival or extinction. To use another vocabulary, these remarks indicate what could also be deemed hope. Darwinism would call it *natural selection*.

The solution of the game of Chicken depends on the signs traded between the players. Truthful cues are conveyed from player to player, but if the cues are not made into representations capable of influencing the opponent before the clash, the game is lost or it ends in collision. The inference concerning the opponent's resolve and determination must be semiotically developed in the mind of the player to whom the sign is addressed. The resolution of a confrontation such as Chicken follows closely the Peircean model of *semiosis*: a sign addresses a receiver as it creates in the receiving mind an equivalent or more developed sign. The set of indexical data that came to Menelaus drove him into producing an inference that transcended the material cues of that specific instant in the race, making him forecast the collision and thus avert it. In conflicts, it is only logical to conclude that signs matter less as types and much more as tokens. While a type is a mere possibility, a token is an actual occurrence that excludes sets of possible occurrences. Now, moving from possibilities to occurrences, one should bear in mind one of Peirce's acute remarks that justifies the central role of indexicality in a comprehensive theory of signs: "An Actual Occurrence always determines a Possibility" (1998: 488). Semiotics of games must give precedence to occurrences, not generalities such as strategies or convention and thus favor the exchanged signs between players as tokens of intended strategies as well manipulations of conventions.

Semiosis among sexual beings ought to favor inflated and costly signs. Inflated signs are not only more noticeable, but they are also more frightening, intimidating, and decisive in a competition. Whether true or deceptive, exaggeration and inflation are inevitable semiotic traits in the competitive universe of sexual beings. Just look at the physical universe around you: and excess is the dominant norm. Inflation and excess happen simultaneously in more than one direction: the *semiosis* of inflated signs occurs among signs competing with other signs that can escalate into greater degrees of excess. Inflated signs are developed into threatening or decisive cues in the mind of the addressee, but – it should not be forgotten – it also indicates the quality of the sender.

Before two youngsters play a canonical game of Chicken with cars heading at full speed toward one another, they can display their intentions with progressively inflated indices. The indices are chosen to intimidate the opponent. Some of the techniques of playing Chicken deserve attention. Kahn (1965: 11) listed them: one of the drivers gets into his car quite drunk, stumbling, with a bottle of whiskey in his hand, taking a swig as he sits to crank the car; an alternative is to wear dark glasses at night to indicate that one cannot see well; to drive blind-folded can also be a good choice of sign; and an extreme and interesting antic would be to pluck out the stirring wheel of vehicle at top speed and throw it through the window. The escalation of signs should intimidate the opponent: all the exhibited signs point to the driver's willingness to clash and pay the price of disaster. This display of signs should work, unless the opponent is oblivious to it.

The signs that the drivers bear before the Chicken game would be effective if they indicated the intention of the player to pay a high cost – to run a serious risk – to win the competition. The decision of the game is directly proportional to the degree of risk announced in each display. That is one step further from the idea that costs determine the reliability of the signal. It is not cost that matters so much in a sign, but how the direction of a sign is shifted toward the player. In that sense, it is not only that signs have a cost, but mainly that the player's strategy would be pursued at all costs. Intentionality and cost blend inseparably.

The multiplicity of equilibrium points in any game of Chicken is reduced because of the potency and the risk involved in the signs delivered: stability is thus achieved in animal and human societies, albeit temporarily and precariously.

Afterword

The radical individualistic stand taken herein is an antidote against the collectivistic and deterministic working hypothesis frequently dominant in cultural semiotics. To embrace methodological individualism in semiotics does not mean at all the complete reduction of sign processes to psychologist postulates, thus denying the cultural dimension of semiotic exchange. It is rather an attempt to place signs where they truly belong: in the micro-level of individuals in relentless competition with one another.

Although with varying capacities, all social individuals are dynamic, active participants in human interaction. They are creative agents, and none of them individually has the power to conduct and determine what happens in social life. The whole comes from the interaction of parts. From the results of widespread games of Chicken social stability is drafted; or to put it differently, from the semiotic interactions of individuals the general culture is made; and it is not the case at all of a culture shaping the individuals within a group. The free individual will pursue his or her own agenda, using all the means possible to vindicate his or her self-interest.

Radical methodological individualism cannot be reduced to methodological psychologism. Taking into account the arguments of the previous chapters, one can see that Freudian theory is not the alternative. The Freudian individual is trapped in the field of the unconscious and therefore unknown drives. The individual is indeed prey to larger forces. That is obvious when Freud discussed the case of psychopathologies. In all of the intellectual variations of Freudian theories, the individual is caught in the web of culture from which it is apparently impossible to find a way out, whether being at the mercy of the contradictory conflict of the pleasure principle against the reality principle, or in the instance of the unconscious force of the *Id* assaulting the *Ego*. That type of individualism is inevitably immersed in historical and biographical determinism.

Yet, the whole picture changes if one adopts axiomatically the assumption that individuals are capable of taking charge of their own self-interest. Such individualism is only superficially akin to methodological psychologism. Even the most selfish individuals are not set apart from other individuals. Their selfishness is precisely established in relation to the pool of other actors. Moreover, it is from self-interest that cooperation emerges in aggregations of originally selfish individuals. In one way or another, culture is at the core of methodological individualism; that is evident if one ponders the emphasis given to strategies in the previous chapters of this book. No holistic theory of signs – no product of the canonical conception of culture that leads to Durkheimean social thought - could ever deal with individuals interacting without cohersion.

Equilibrium does not come first to control and stabilize individuals. It is the other way round: selfish individuals interact and then order is precariously formed. What Claude Lévi-Strauss (1949) proclaimed in his classical study on the elementary structures of kinship must be completely rejected. Culture is not a synthesis that transforms and allows the overcoming of the pressures of the biological. To escape fully from the pressures of biological life is unattainable; in fact, culture cannot work against biology. It is always nurtured through nature (Ridley 2004). Biological strictures are the threshold of any living organism. Cultures must operate from what is biologically given. For this reason, the present argument is that the most basic ground of sign production is not the social group, but biological mechanisms, and among them precedence must be given to sexual selection.

To validate completely this contention, it is necessary to provide an answer to a potential objection: Can an interpretative model strictly based on individualistic competition explain group phenomenon such as war between cultures? Is war the consequence of group pressure or does it come from the pressure of sex? Therefore, to demonstrate that it is intellectually productive to hold a methodologically individualistic stand on the issue, and before presenting concluding remarks, the next sections in the conclusion of this book will examine if cultural causes are the sole means to comprehend – for instance – the events that Thucydides describes in his *History of the Peloponnesian War*. Is it mandatory to opt for conventionalist assumptions to understand this particular moment in Greek history? Is cultural difference or sexual selection the cause of group conflict? Providing an answer to this question will redefine the role of the semiotics of culture during one of the most important and devastating conflicts of the ancient world. The limitations of conventionalist semiotics will be revealed.

The cause of conflict between cultures

Some questions, both historical and theoretical, about the function that cultures and their conventions presumably had not only in the Peloponnesian War, but also in other aspects and practices of societies must be considered first:

1. Despite being considered by Thucydides' narrative one of the elements of the Peloponnesian conflict, is cultural difference responsible for the origin of the war? If so, in what way?
2. If not, is cultural difference just a source of arguments supporting and justifying military strategies?
3. Did Thucydides conceive cultural difference as the exclusive cause of the conflict or just one of its factors?
4. Was the Peloponnesian War, like so many historical events, a mutant occurrence that resulted from multiple causes?
5. What general theoretical consequences can one infer from the examination of the role of Pericles' and Thucydides' idea about culture in the first years of the war of Athens with Sparta and its allies?

The three speeches of Pericles in *The History of the Peloponnesian War* display the primordial Western reflection on the role of culture both in war and in society. Among the public communications by the one whom Thucydides regarded to be the voice of all Athenians, the funeral oration substantiates the city's military superiority over Sparta as a direct effect of her culture. Democratic culture is a martial asset, for in Athens the constant practice of liberty (*eleutheria*) makes of her citizens not only better soldiers, but also individuals committed to the decisions of their assembly. Although other military factors and elements should be taken into account, it is indeed through the interaction in the public sphere that the citizens come to realize the best way to fight (Hanson 1991: 43). Therefore, following Pericles' vision, the link between culture and conflict is acknowledged. If the link is valid, cultural order had an active and formative role in the struggle for power in the Greek world.

From a theoretical viewpoint, and more important, because of the antiquity of the document that is Thucydides' *History*, the conception of culture presented through Pericles' speech should be considered indisputable and timeless, as anthropologists have dedicated themselves to demonstrate because it cuts across the most varied historical periods and circumstances. In both war and peace, the notion of culture put forward in the funeral oration matches modern definitions of cultures as conventional systems. In other words, the cultural system is an *a priori* social factor; it presents values, norms, and procedures capable of guiding, determining, and ordering behavior within the common arena of society. Ordering implies two meanings: it is what fashions the unity of socialized beings and what commands their actions.

In conventionalist terms, the culture in which one is born and educated is inescapable; social actors experience it at a moment when they are blank slates. The individuals are then susceptible to the determination of traits that social cultivation imposes upon them. The marks of socialization cannot be erased and they make up the common heritage of a group, being what distinguishes a group from other social aggregations. Again, the logical inference of this conception is that cultural traits build up the radical distinction between Athens and Sparta, and the consequent conclusion is that, eventually, conflict has to happen.

Here one finds a hypothetical explanation for the origin of the war that spread itself relentlessly throughout the Hellenic world during the fifth century B.C.E. The stability and the permanence of cultures should answer the puzzle why the war lasted for so long and why the individuals who commanded it were powerless to find a way out of the impasse that would be unyielding and eventually cause large devastation. However convincing the cultural account may be, and therefore dependent upon a radical conception of culture,[1] it is unfortunately contradicted by Thucydides himself, who also claimed that the historical contingency, not the cultural structure of the groups, was the original motive for the emergence of the conflict. He clearly stated: "what made the war inevitable was the growth of Athenian power and the fear that this caused in Sparta" (1. 23).

Still, one should not take Thucydides' words as the definitive causal explanation of the Peloponnesian War. Some contemporary historians doubt if this interpretation is the complete motive for what happened. Thucydides formulated this supposition based on a kind of collective psychology, a direct transposition of his generalized assumptions about human nature. An unambiguous historical analysis of the war requires another reply to the question.

The presumption of Spartan fear of Athenian imperial growth is unwarranted. Kagan (1969: 346) notices that, between the years of 445 and 435 B.C.E., thus for a decade, Athens had not made substantial additions to her imperial power. This means that other factors such as pride and care for the city's reputation should be taken into account to explain the conflict. Warfare was indeed among the Greeks and many other social groups a practice that could establish the value of both individuals and groups. Thucydides himself presents a far more powerful hypothesis in that direction, declaring that the war resulted from a triad of motives: honor, security, and selfish advantages.

The triad of establishment of prestige, search for security, and the fulfillment of selfish interests allows for a more complex understanding of the multiple causes of the conflict. It implies that there was not a single and general cause but – as in the game of *Chicken* – a series of local and varied motives that gradually grew into a confrontation of larger proportions. It explains why a civil war in a city of minor importance in the Greek world, Epidamnus, could tow Athens and Sparta into a devastating conflict. Quite true, the civil war in Epidamnus has to be considered a marginal event, but the system of alliances between Greek city-states favored the uncontrolled upsurge of conflicts, involving the great power centers of the time that could not allow to be challenged. To be indifferent to challenge was just a mask for weakness and cowardice. In this way, the dispute between Corcyra and Corinth was transformed into the clash between Athens and Sparta. It is accurate that the historical grievances and rivalries and the different cultural assumptions of Athens and Sparta fueled the conflict, but that could not be all.

In *The Outbreak of the Peloponnesian War*, Kagan observes that, in the deliberations before sending the ultimatum to Athens, Sparta was divided about the conflict. As always, the belligerent spirit of the Spartans was far from impetuous. They feared constantly the rioting of the helots, a population of Sparta's slaves. The helots hated the Spartans, who had reduced them to slavery. It is easy to see that the helots, being more numerous than their masters, could take advantage of the war with Athens and attempt an uprising.

Both externally and internally, the power balance in Sparta was far more flimsy than one can superficially presume. Externally, its allies like Thebes and Corinth acted freely without obeying the city that commanded their league. There were no guidelines in the Peloponnesian League and the same happened in the Delian League under Athens's leadership. The alliances between the cities were mainly defensive. The war checkerboard was a kaleidoscope, reshaped endlessly under the pressure of singular and local interests, frequently contradictory.

Mytilene, for example, despite being under oligarchic rule, was part of the democratic league; but the Mytileneans tried to take advantage of Athens's weakness to break with the empire and then incorporate other cities of the island of Lesbos. In the peaceful period, before the beginning of the hostilities between Athens and Sparta, the Mytheleneans tried to gather support from the rival Peloponnesian League; they knew that the Athenians would not tolerate their expansionist plan. At first, not recognizing any gain in the Mytheleniean proposal, the Spartans refused the idea, but, as the war raged on, an uprising within the Delian League became an attractive alternative; the Spartans then encouraged it. Everything would be in synchrony: Sparta would invade Attica in 427 B.C.E. while the Delian League would be fractured by the Mythelenian revolt. The Spartans hoped that the insurrection would spread itself through more than one city, weakening Athens even more as the Athenians were going through the devastation of the plague and the depletion of their public finances.

For everyone involved in the process, the steadiness of forces as the war developed was increasingly subjected to the whimsicalities of circumstantial interests. One just has to think of the many betrayals, disloyalties, and breaches of agreement that happened throughout the conflict. Not even Sparta, a city most clearly ruled by an *ethos* of obedience, was immune to domestic instability. Sparta constantly went through moments when no one seemed to take charge of and responsibility for the conduction of public policies; power was split continually between her two kings and the *ephors* (five overseers and guardians of the collective interest, elected annually without the possibility of reelection). Sparta's decision to trigger the war came in such a political void. A passage in Thucydides (1. 80-86) describes this moment: it came after Archidamus had advised them against the conflict with Athens, asking the assembly to take some time from the hostilities to be better prepared for the long haul. Then, one *ephor*, Sthenelaidas, rose and aggressively argued against the cautious advice of the king, throwing Sparta into warfare. Sthenelaidas played *Chicken* against Archidamus.

In its process of expansion, the Athenian imperialism was not without contradictions and paradoxes. Athens, the city of democratic liberty, reduced the freedom of other cities; but, if the Athenians had adopted a policy, unthinkable for Pericles, of denying their imperial destiny, the decision would have led them to face external and internal threats. Externally, Athens would live under the menace of being coveted by other cities. Without the imperialistic expansion, Athens would be under the constant threat of internal rupture and dissatisfaction; scarcity and deprivation would be the common fate of the Athenians. How could the city refuse the abundance and the riches that the imperial power propitiated?

Life in Athens was never free from the fear of rupture. Disruptive divisions are hovering presences in democracies, for the unity within a democracy is always precarious; its harmony depends on the consensual discussion of conflicting interests. Thus, in his memorabilia of Socrates, Xenophon (3.5) recognizes in the Athenian democracy a permanent tension among individuals who deem themselves brothers and yet are persistently dedicated to internal disputes of Chicken. Unyielding belligerence is the Athenian rule. More recently, Loreaux (1986: 198) underlined what Classical historians like M. I. Finley pointed out: as the history of democracy unfolded, in the Greek lexicon, the term *stasis* changed from meaning "political position" to "sedition."

None of the circumstantial causes presented so far can be considered a cultural explanation for the Peloponnesian clash. If that is the case, why did the war follow a pattern that was relatively stable? One way of explaining this is to admit that the stability of the impasse between Athens and Sparta is an effect of the bipolarity of the struggle coupled with the lack of definition of military superiority on each side; but this has nothing to do with the cultural differences between the cities. One should try not to see too much in the struggle. The impression of a point of equilibrium comes from the bipolarity of the leagues around which the conflict unfolded.

Even if the war took startling turns because of action of the other independent Greek cities, everything returns to the polarization of Athens and Sparta. Thus, the cultural justification of the war develops into some other explanation: cultural factors are really no more than elements of argumentative strategies.

If cultural factors were strong enough to conduct the conflict, the Athenians would have never disregarded Pericles' propositions. For the Athenians what Pericles wanted was to create a strategic harmony during that moment of the war. Whatever he proposed had no special weight for invoking the cultural principles of the city. Pericles' speeches presented arguments in favor of a collective strategy with one weak point: to accept for the moment an orchestrated strategy was feasible, but to support it for a longer period of time was much more difficult. For that to occur, the maintenance of an unremitting contract would have meant the entire reconfiguration of Athenian notions of social competition. Paradoxically, Pericles' cultural arguments demanded that Athens become another city with another culture. Was that possible? According to conventionalist and oversocialized conceptions of culture, it was not. But is conventionalism right? Or is there a different way of conceiving culture?

Sexualized culture

Contrary to what conventionalism maintains, cultures are not forces of impersonal coercion. They are stages for social actors to compete with one another. Social prestige and group recognition translate into the individual prospect of being selected sexually. From the viewpoint of the individual, cultures do not even out; they establish the condition for the recognition of personal differences. The same happens when comparing cultures. The distinctions among individuals still stand out, but these are different kinds of exclusion. Cultural differences make individuals of other groups small and childishly inept. Their actions and messages are utterly awkward, so chances of being selected sexually shrink. Once more, the mechanisms of cultural differentiation act to single out who is biologically excellent, not with the purpose of segregating some, as is the case within a culture, but to reject all who come from the outside. Cultural performance becomes a fitness indicator.

These ideas follow Darwinian contentions presented in *The Descent of Man; and Selection in Relation to Sex* (1998b), where in 1871 Darwin conceived of sexual selection as a universal biological trait of organisms that have had their sexes separated. Sexual differences are shown in the primary sexual character of reproduction such as organs but also in secondary sexual characteristics. The asymmetric development of physical traits in males and females occurs for a sexual reason, albeit apparently unconnected with primary sexual reproduction. They are "the greater size, strength, and pugnacity of the male, his weapons of offense or means of defense against his rivals, his gaudy coloring, and various ornaments, his power of song, and other such characters" (Darwin 1998b: 215).

Sexual selection is a force that precedes the advent of human cultures. Therefore, the strength of sexual selection is greater than any posterior set of conventions. In Darwinian terms, the sexual splitting of organisms into males and females creates a division of attitudes, behaviors, and functions that are mirrored culturally. Females select for reproduction the males with evident biological qualities. The certification of the masculine qualities demands proof, indications that truly represent individual excellence. The indicators should be shown in the individual, who ought to be capable of paying a higher price than his competitors for his actions. To survive the extreme dangers of a war in which the other competitors perish is certainly a mark of individual biological fitness.

For the Greeks, the connection of military belligerence and sex was clear enough.[2] In 413 B.C.E., at the end of the Peloponnesian war, Aristophanes'play *Lysistrata* proposes comically the solution to put an end to the interminable struggle. The idea is that this will happen if the women of both cities refuse to have sex with their males. In more than one scene in *Lysistrata* is obvious the presentation of a detail representing how cultural distinctions serve the purpose of sexual selection. It can be seen when the Athenian Lysistrata talks to the Spartan Lampito[3] to forge an agreement between the women of each city, as well as when the Spartan ambassadors come to seal the peace treatise that will appease the sexes. Even if the Greek language was common to the conflicting sides, Aristophanes depicts the dialogues of Lampito and the speeches of the ambassadors as odd and rustic gibber, as the Athenian audience used to perceive the Doric dialect of the Sparta. This is more than sheer cultural prejudice. For the Athenians, the linguistic expression of the Spartans sounded foolish, gauche, incompetent, and therefore without great sexual attraction.

Sexual selection also explains the turn in the Spartan assembly that declared war to Athens. The old king Archidamus II would have no special reason to demonstrate his sexual fitness. He could be cautious, for he had surely made known his excellence in the past. The impetuous and young *ephor* Sthenelaides demanded war, claiming the urgency of preserving Spartan honor.

The theoretical explanation for the phenomenon is as follows. The bewildering diversity of life forms and human cultures come from the competition spread in nature through the selective pressure of sex. Within a population, males compete with other males to be selected sexually. Another kind of competition occurs between the sexes. Sexual cooperation and mutualism are preceded by a conflict of interests separating males and females. The reproductive success of males does not depend upon elaborate and extremely strict selectivity. Male organisms can beget a greater number of descendents sequentially. Females, in contrast, have a costly reproduction: once pregnant, their reproductive capacity is put on hold until well after the offspring is born. To reproduce, the female organism must wait, while, at least in theory, the male can go on spreading his genes throughout the population.

Putting male indiscrimination against the extreme selection of females, one can see that reproductive life is an evolutionary arms race. Males try to find new means of seduction and persuasion, while females will tone down the suspiciously deceitful male strategies. It is inevitable that females will check, evaluate, and refuse critically what is offered them as a reproductive option. Females tend to reduce indiscriminate mating, which is in the exclusive interest of males.

Females provoke exclusions of two kinds: within the group, for in the generally non-monogamist natural world many males will not be chosen to mate; but also rejecting individuals that are nor part of their population, for whoever is not from the group will have even fewer chances to procreate. As time goes by, the population will split, creating another group. This answers for the phenomenon of biodiversity and cultural differences. Males will search for alternative niches and will form other social units, but the principle of selection remains the same: inside the new group, the male better adapted to female demands will have an advantage over outsiders (Tregenza 2003: 929). Competition is the fate of sexual beings.

If sexual beings cannot suppress completely the intense competition of individuals, whether internal or external, social experience is never totally stable. A volatile and precarious equilibrium rules over life among sexual organisms. For a leader orchestrating social aggregations, enforcing collectivistic policies is much more necessary than giving speeches, however moving and persuasive they may be. Even if a viewpoint has been temporarily accepted, the group is always under the threat of dispersion, fueled by incessant conflicts of interest. Unsurprisingly, Pericles' speeches during the Peloponnesian War were the object of refusal and resistance from his fellow citizens.

With all this in mind, one can easily comprehend how biological diversity and speciations as well as cultural differences were possibly created. The cause of cultural diversity is the generalized and spread-out competition conducting sexual selection. But there is something more fundamental than cultural differences setting groups against one another. Males who, in Darwin's words (1998b: 632), woo females are always biologically equipped with weapons to fight their rivals. To be better armed, to be more successful in fighting, is to be more attractive. Sexual selection of individuals creates the emergence of irredeemable cultural differences. Cultures are not the origin of group conflicts; they furnish strategies and potentially convincing arguments for the struggle.

The traditional fallacies of cultural semiotics

One of most fundamental mistakes that the conventionalist conception of culture disseminates is the belief in the overwhelming role of conventions in social life. The strength of this notion comes from its distinguished intellectual tradition that goes from Thucydides' interpretation of the Peloponnesian War to Plato's hope of coordinating the ideal society, landing in Augustine's assertion that the will of individuals are subjected to a contract with God under he fearful penalty of eternal damnation. As stated before, the powerful drive of Thucydides, Plato, and Augustine ripple silently through modern anthropological descriptions of signs exchanged in culture settings.

The canonical idea of culture presumes that social actors and the signs that they communicate are deeply determined by the prevailing assumptions of their groups. This is the guiding spirit of contemporary cultural studies. Then, signs circulating among individuals are made into ghostly shadows of conventions; when, in fact, any representation not only describes events, but it can primarily indicate the intention and the purposes of the transmitting individual. If that is so, the group must play a secondary – although undeniable – role in the interaction of social actors. However, that is not the only mistake that conventionalist semiotics propagates. Below are underscored some of the governing fallacies in the humanities and the social sciences that can be avoided if one denies priority to conventions in semiotics:

1. *The fallacy of the separation between the human and the natural.* This fallacy is a major obstacle hindering the social sciences and the humanities. It is impossible to uphold it as biology shows increasingly that life is a radical unity, a blend of what was conceived as separate for so long
2. *The fallacy that human cultures are barriers between groups, generated through learning, which is always necessary to the socialization of human beings*. The idea is unacceptable, not because human groups are not distinct from one another (who could deny that?), but because the distinctions of societies are just erections of genetic and sexual hedges. The motive for cultural differences is strictly biological, and it is connected with the fact that males and females are guided by non-convergent biological interest emerging from their reproductive anatomy, and not from their culturally learned gender roles. As a consequence of this biological interaction, the process of sexual selection favors some males to the detriment of others, who will search for other females, thus creating or moving into other groups of sexual organisms. Cultures then emerge, and with them the fact that common mechanisms are at work in nature and human cultures. Different cultural socializa-

tion is not an end in itself, like cultural anthropology suggests, but an instrument of the process of sexual selection. Its origin antedates human and animal cultures as well as any process of social learning: it is common to all sexual beings, and it acts predominantly at the level of the individuals.

3. *The fallacy that cultures determine the action of individuals – and by extension their exchange of signs – within groups.* Such a conception is essential for the two modes of identity in cultural studies: the identity of the group and the identity of individuals. Group identity is closely dependent on the assumption that groups are stable totalities. The problem with this conception is that in life everything and everyone is under relentless evolutionary transformation, whether at a faster or a slower pace. Cultures change internally and externally. Cultural practices of other groups can be absorbed with greater or lesser difficulty. No group is completely sealed off from other groups. No language is totally impossible to translate into other languages. Cultures transform, amplify, modify, or conserve other group's cultural products; and such cultural encounters do not necessarily breed violence; they are moments of critical evaluation for the culture that meets what is not part of its own structure. If permanent innovation is an efficient strategy that sex grants to so many species and living organisms, it is an unwarranted distortion to favor continuity, stability, and fixity in cultural processes. Individuals in ceaseless interactions with other individuals must be capable of adopting and inventing multiple strategies – stability and fixity are obviously negative qualities in this case – if not, the price is either exploitation or extinction.
4. *The fallacy of collectivism.* Collectivism and conventionalism are infantile diseases of the humanities and the social sciences. Collectivism is a sheer abstraction of the interpreters of social groups. Look at clusters of individuals and answer this question: Where is collectivity? The collective is inevitably a fuzzy entity, inferred ethereally from the actions of self-interested individuals. A collateral effect of this theoretical delirium is the mystification of presumably all-powerful, all-determining groups. As it is clearly visible in Thucydides' rendering of Pericles' funeral oration, collectivity is less a reality than an argument, often used to attempt at coordinating the action of individuals, if not to justify oppressive social practices. While untenable as a theoretical tool, collectivism is easily turned into pious and moralizing arguments; such an assumption gives to collective wisdom the sacred aura of righteousness, when group authoritarian-ism is repeatedly the cruelest and

coldest of monsters. The utmost development of this absurdity is the sentence: "vox populi, vox Dei," so many times flaunted as unquestionable truth.

5. *The fallacy of the force of the group proportional to its extension that gives to mass culture the role of an overwhelming cultural whole with greater and greater power.* Indeed, it is quite the opposite. The larger the group, the greater the chance the individual has of escaping the control of other individuals (Olson 1968.) In larger groups, it is reasonable to expect more free riders taking advantage of other individuals. For that reason, in history, larger groups develop specials forces that police and enforce the collective rules, which are always on the verge of being shattered. For group control to be truly effective, two basic conditions must be met. First, the community must be small cells, at least not extensive enough to allow the rise of free riders. If the group is expansive and not contractive, individuals can avoid facing the outcome of games of Chicken that are best played on a dual basis, although Chicken can be also the mobbing of a sub-group against the deviant and disobedient player. In Augustine's time, it must be noticed, the effectiveness of the Christian community – the city of God – over the brethren was directly related to their relatively limited size. Among North Africans, even the ones who deemed themselves followers of Christ, Catholics were a self-enclosed minority. Second, to complement the action of members of a small group relentlessly playing Chicken among themselves, the threat of the most powerful enforcer, God Himself, omniscient and omnipresent, was a deterrent capable of framing any wicked social actor.

Altogether, it is undeniable that sex fosters complexity. Sexual beings have complex individual constitution as well as complex modes of association and sociability. Without association, sexual beings do not reproduce. As discussed in the chapter "Sex, Signal," if an organism is sexual, it carries in its genetic make-up other lineages, thus creating a web of widespread and progressive relations. Such a progressive intricacy of relations should be cumbersome if there were no reduction of complexity.

The cost of complexity is reduced through the creation of general and encompassing rules that minimize the expense involved in remembering what is appropriate, correct, and advantageous in ever-growing webs of individuals relating to other individuals. Conventions are nothing but after-effects of interlocking sexual relations.

At this point, it is important to realize that the efficiency of rules is directly proportional to the individual's capacity of generalization. The capacity to generalize is not a direct product of the inductive abstraction of perceived social relations. Even before existing in society, human individuals are naturally limited in their potential to generalize by the evolution of their brains that, through critical assessments, and since the earliest moments of life, exclude and segregate what is inconstant, while what is constant slowly acquires the strong stature of a cultural rule, a rule to follow in interactions with others. Why would children not only tolerate, but also thrive and show eagerness for repetition and redundancy if not because of their urge to grasp social rules?

The future of cultural semiotics

All that should lead to the belief that cultural semiotics in the near future will profit tremendously from upcoming studies on the neural mechanisms of the human brain. Although this area of research is expanding into uncharted territories, some of the evidence of empirical investigation is already fruitful for cultural semiotics.

An astonishing revelation of neuroscience is that the brain has no unifying center. The brains of individuals receive data through the eyes; the data then travels through neural pathways with distinct and specific functions. One pathway is filogenetically ancient, while the other is newer and present in developed primates and thus human beings. A clear division of labor is at work in the systems of the brain (Ramachandran and Blakeslee 1998: 73). The older system allows navigation in the environment, while the newer one participates actively in the identification of what is in front of the organism. One system indicates how the organism should act, while the other establishes the identity of other organisms and things in the natural world. Division of labor is so central for the functioning of the brain that each neural pathway and its cells are highly specialized in recognizing distinct visual attributes. Such splitting of functions optimizes the human brain as a highly efficient social and cultural tool.

The brain is therefore not inductive; the dominant mechanism in the brain is deduction, for it operates from basic general representations. From general representations, signs are referred gradually to individual objects. As a result of the division of labor, the movement of brain activities is always from generalities to particularities, and not from the consideration of particularities leading to the mining of general qualities.

Indeed, a noun designates at first not a singular object, but a class; when the noun is coupled with other words, it indicates individual things. The word "dog" coupled with "this" and the predicate "white" becomes an expression capable of describing individual white dogs.

The process is both linguistic and visual. The visual brain recognizes first the quality "red" because it has cells specialized in the identification of colors, and then, the specific forms of red; but yet, for this reason, it is possible to close one's eye and evoke "redness," confirming what Charles Pierce's system of categories called firstness: a generalized quality present in the mind of viewer, an ontological possibility not yet embodied in singular and concrete objects.

In this scenario, it is impossible to demonstrate that the brain has a center with the function of unifying sensory data that can be called consciousness. If individual consciousness has been hard to localize, collective consciousness is an even more arbitrary supposition. Such collective consciousness is just a mirage of the human capacity to generalize. It may well be that the persistent and unshakeable identity of the individual players in the previous sections dealing with game situations is just a convenient assumption to clarify the moves and the goals of strategies. Examined closely, the idea of a stable self seems to rest on a strange sense of emptiness. What is assumed a self with perfect and lasting identity is just a bundle of conflicting perceptions with an occasional equilibrium. Like no other philosopher, David Hume in his *Treatise of Human Nature* (Book 1, Section VI) defined the self in terms that are quite close to the description of contemporary brain science. For him, the self is nothing but "a collection of different perceptions, which succeed each other with an inconceivable rapidity, and are in a perpetual flux and movement."

The only thing that is certain is that the brain functions through parallel processing. Nothing is passive in the brain. All the brain does is generate an active and primarily hypothetical representative construction. The brain judges stimuli following a deductive mechanism, identifying what is and is not regular, shedding away false clues, only then accepting the validity of the representation.

All representations are originally hypothetical generalizations. As parallel processing rules mental activities, the brain deems irrelevant gaps in the parallel processing of the external data. Persistently and actively, the brain assesses external data, operating with inherited mechanisms of judgment that precede and allow any mode of cultural learning. The initial movement of the brain is to identify a faulty hypothesis, then, after discarding the potential mistake, the brain builds up what is accepted as a truthful representation. The self and the perceived world are not stable and fixed: their stability is just a precarious equilibrium. This complex process goes unnoticed because of the extraordinary speed of neural time (neural time is measured in milliseconds – one thousandth of a second – and microseconds – one thousandth of a millisecond.)

Also noteworthy is the overwhelming presence of sexual mechanisms in the brain. The brain is the seat of pleasure and beauty (as well as pain and revulsion). In the brain is also located the amygadala: the threshold of any emotion. Brains – preferably large brains – are necessary for the development of complex strategies of seduction; they are vital for the successful mating of individual organisms. Because the brain is closely associated not just with pure reason and logical inferences, but also with actions and sex, it should not be any surprise that the human brain has been under the pressures of natural and sexual selection, evolving incrementally for at least the last 60,000 years, and probably for as long as human life exists.

Finally, the contention that the human mind is a blank page, devoid of evolutionary origin, and consequently notoriously submissive to what it receives, fosters the illusion of the power of conventions. Players can develop strategies because they carry the capacity to generalize their actions into a plan that may be successful in the competition with other individuals. Individual players, attuned to their self-interest, can then develop strategies according to their biological capacity and considering the actions of other players. That is the primary goal of the semiotic interaction of players within a culture. Only then can a generalized cultural conception emerge. The conceptual insight of game theory is a valuable asset for a semiotics of culture conscious of the actions of individual players in cultural arenas. Collective representations can do much less than cultural conventionalists have systematically presumed.

Notes

1. The term *total ideology* was borrowed and transformed from Karl Popper (1963: 212–223). In his critique of Hegel and Marx, Popper called total ideologies the belief that the social habitat of a thinker, and by extension that of any social actor determined the theories and the opinions held by them.
2. Another trait of Freudian total ideology reveals itself in the consideration of what happens when the analyst faces the phenomenon of *Verneinung*, which can be translated as denial. Freudian psychology claims that, during a therapeutic session when a patient denies the validity of a psychoanalytic interpretation, the denial cannot be admitted as a simple and straightforward "no." It is not appropriate to sustain that the reactions and the criticism of patient will be all the time neutral or even correct. Both the therapist and the patient can be right or wrong. It would be healthier to recognize the common possibility of errors. It is thus astonishing that the therapist is granted an aura of unquestioned authority, albeit provisionally, and that any counter-argument be silence and labeled as resistance. To the patient is left only the option of adopting the proposed interpretation or leaving therapy. That would paradoxically confirm the denial, making more and more difficult any refutation of the therapist. Freudian theory can be as totalitarian as Marxist authoritarianism. Freudianism and Marxism share the same kind of holistic, historicist, and conventionalist foundations. Another mode of cultural determinism is in Lacan's emphasis on the cultural importance of what he calls *Le Nom du Père* – The Name of the Father. The Lacanian description of Le Nom du Père is as follows: "*c'est dans le nom du père qu'il nous faut reconnaître le support de la fonction symbolique qui, depuis l'orée des temps historiques, identifie sa personne à la figure de la loi*" ("it is in the name of father that we must recognize the support of the symbolic unction which, from the dawn of history, has identified his person with the figure of the law," Lacan 1976: 77).
3. What Saussure really thought, what route he would have envisioned for linguistics, is outside the scope of this book. *Communication Games* is concerned with the impact of Saussurian ideas on social semiotics; and if that is the case, the responses to the edition, distorted or not, of Saus-

sure's course on general linguistics put together by Bally and Sechehaye, are what is valuable. Bally and Sechehaye's edition was the inspiration of Barthes's *Elements de Sémiologie* (1964), and the reflections of Lévi-Strauss (1945) on linguistic and anthropology. For interpretations of the *real* Saussure, see Harris (2000, 2001), as well as Kroener (1973). Robert Godel's (1957) early presentation of the manuscript sources of Saussure is still of great use; see also Bouissac's (2004) analysis of the Saussirian influence.

4. Werner Jaeger's reaction to the universal expansion of the idea of culture to groups other than the Greeks is interesting to read. He rejected the trend of applying the idea of culture to the modes of life of any other nation. This application betrayed the main function of *paideia*, which is to offer an ideal goal to citizens. Jaeger (1939: xvii) asserted that in a vague way one could talk of "Chinese, Babylonian, Jewish, or Egyptian culture, although none of these nations has a word or an ideal which corresponds to real culture."
5. Because *Communication Games* is not an overtly political book, in no way should it be understood as a justification of conservative *laissez-faire*. If the individual is radically and properly placed at the core of social and political concerns, one must conclude that personal freedom and democracy – fallible as it may be – are preferable to collectivistic social options. The role of the government is to protect individuals in their personal choice, diversity, and singularity. It is unacceptable and politically unwise to leave the issue of individualism in the hands of the political and social right wing. The uncreative stance of conservative politics and its collectivist emphasis on patriotism contradict a radical individualistic philosophy. Obviously, the left needs a renewed set of concepts, as Peter Singer observed in *A Darwinian Left* (1999).

Part 1
Canonical games

1. Conflict

1. Herodotus (7. 144) gives credit to Themistocles for shifting the Athenian military force toward the navy. Themistocles persuaded the Athenians to use the surpluses of public funds, extracted from the discovery of silver mines in Laurium, to commission the building of 200 triremes. The ships were to be employed against the Aegenians, but the war with the Aegenians never came. According to Adcock and Moseley's (1975:

26) appraisal, from 493 to 483 B.C.E., the Athenian fleet grew from 40 to 400 triremes.

2. Pericles and Archidamus were *xenoi*, which meant something more than friends. The literal translation of the Greek term, *xenos*, is "guest-friend," and refers to mutual and ritualized obligations of hospitality linking them. It was an ancient institution among the Greeks: "The root meaning of xenos was 'stranger,' 'outsider,' 'foreigner'; and *xenoi* in the sense of 'guest-friends' were always foreigners, that is, members of two different communities" (Cartledge 2003: 159.) *Xenoi* were bonded by trust that extended beyond each "guest-friend" and connected their offspring.
3. That is not the only appraisal of the *epitaphios*. More recently, Griffin (2003: 62) considered Pericles's funeral oration an actual portrait of "the unique quality of democratic Athens, its openness, its freedom, and its cultural achievements."
4. Athenian society had more than one way of granting social merit. Behavior in the battlefield was certainly one of them, but there were others. To be first in a competitive dispute, in the gymnasium, in athletic contests, in the theatre, in social hierarchy was a pervasive ideal (Plecket 1996: 510). Without forgetting that rhetorical disputes were the foundation of democracy, one can deem Athens an agonistic society. The goal of the individual was not just to win, but also to excel above others who were his equal. All across the Greek world, the citizens dreamt of fulfilling the ideals of *agon* and *stasis*. Politically, *stasis* implied internal strife, and it developed from tyrannical take-over to ritualized, diffused democratic disputes. In this context, democratic traits developed, and they meant regulation of internal political conflicts, the constitution of civil unity, and the cultivation of public communication (Gehrke 1996: 464.) The whole point was to create social order from the conflicting stance of individuals; it is no wonder that so many Greeks admired the internal cohesion of Sparta.
5. Solomon (1985) thinks that, about contagion, Poole and Holladay exaggerate Thucydides' medical wisdom. It is true that Athenian medical literature ignored how contagion spreads a disease, but the Athenian populace had perceived the phenomenon; they abandoned an established cultural rule and avoided taking care of the sick. That is astonishing for a group so concerned with proper ritual burials, as were the Greeks. Moreover, at that moment, the invasion of the city by enemy forces must have been frightening and apparently imminent. The Athenians were trapped. If the invasion were not forthcoming, the citizens

would have run from the plague, which lasted about four years, leaving Athens behind.

6. The hoplite went to battle wearing a helmet that weighed about five pounds or more. The helmet's design, without holes for the ears, covered the head and the collarbone, enclosing almost all of his sense organs: the ears, the mouth, the eyes, and the nose. The hoplite was left only with his touch to orient himself during the conflict. Hanson (1989: 71) realizes, with acumen, that hoplite fighting must have been a frightful experience, not just because of its bloody slaughter, but also because of the terrifying sense of isolation: "The isolation created by the helmet determined that each individual seek close association with his peers." Unable to hear and crowded into the mass of his companions, the hoplite only knew what was happening by feeling the pressure of the individual bodies around him. Ignorance and isolation increased the alarm, the horror of the pitched battle, for "each man withdrew into his own shell of fear" (Hanson 1989: 100.)
7. Sahlins's (2004) critique of Thucydides ignores that the Greek historian was cautious enough to emphasize the role of human nature and individuals during the Pelopponesian conflict – most clearly the initial role of Pericles and how his death left Athens rudderless – without neglecting to leave a record of Pericles' interpretation of the war, at least in the Funeral Oration, that relied so much on the antagonism of Athenian and Spartan cultures.

2. Coordination

1. The difficulty with *Cratylus* begins with where to place the dialogue in the whole bulk of Plato's works. Some interpreters and historians of philosophy argued that *Cratylus* should have a place along side the early dialogues, among the "Socratic" texts, concerned with telling the events of the historical Socrates' life. At this point, Plato had not yet interfered with the narrative of facts, and introduced his theories. Considering that Plato's theories normally show up in the middle and later phase of the philosopher's works, Kraut (1992) notes that if *Cratylus* has its own share of philosophically abstract arguments, it has to be a work of Plato's middle years. Luce (1965) argues that Cratylus was composed after Plato's shorter dialogues, showing Socrates' greater influence; therefore, it comes before *Phaedrus* and *The Republic*. Cratyllus had to be part of Plato's middle years. Brandwood (1992) is positive that Cratylus is more Platonic than Socratic. It should be near

Euthydemus, *Phaedo*, *Theatetus*, and *The Sophist*. MacKenzie (1986: 124) feels that the work was written at the final period of Plato's philosophical activity. To make things even more complicated, Shorey (1986: 75) affirms that *Cratylus* is a kind of abridged inventory of Platonic thoughts and classifications.

2. Ketchum (1979) considers Hermogenes' conventionalism naïve, as it cannot give the inevitable step forward. It does not perceive how agreement and socialization are necessary to confer meaning to a name. Meaning is not an attribute of names granted by arbitrary, particular, and individual decisions. Meaning and naming are public.
3. Hermogenes' thought is so tortuous and inept that Weingartner (1970) suggests that in this dialogue Plato had in mind more than a mere criticism of relativistic theories originated in Protagoras. Plato would be criticizing also the negative effects of Athenian education. Weingartner's interpretation can be supported, if one bears in mind that Cratylus, a younger contemporary of Socrates, was Plato's first teacher. From him, Plato learned that the world of senses is in a constant flux, and from that, he inferred that there ought to be a world with opposite traits. Fluidity entails necessarily fixity, from which it surfaces. The other world has to be permanent, fixed, and to know is to know it. In *Cratylus*, Plato was reviewing critically his own education.
4. The Platonic model of discovery depends on a face-to-face interaction. The living discourse is a condition to arrive at the truth. For Plato, the written dialogue is a hopelessly dead expression. Writing does not allow the choice of interlocutors. As an indication of Plato's aristocratic view of knowledge, he dismisses writing because it can come to the improper hands, the ones who would not understand what is written, or who have no interest in the probing for truth (*Phaedrus* 275, d-e). McComskey (2002) criticizes Plato's critique of rhetoric. The Platonic stand in favour of a timeless truth is in direct opposition to Gorgias' epistemological stand. Gorgias' rhetoric is based on Kairós, the right moment, the right situation, and that can occur at any time.

3. Contract

1. Casamassa (1952) reported that in 1945 two boys of the Youth Association of the Church of Saint Aurea, when digging holes to build a basketball court, found a slab of marble with the funerary eulogy of Augustine's mother. Although extensive excavation followed, the search for the missing part was unsuccessful. The gaps in the sentences,

which were originally written by the Consular Anicius Auchenus Bassus, could be easily restored, for they had been copied in ancient manuscripts. The parts of the following sentences in capitals were inscribed in the marble slab; the lower cases complete the inscription: HIC POSVIT CIN(eres genetrix castissima prolis)/AVGVSTINE TVI (altera lux merit) or TVI(s altera lux meritis)/QVI SERVANS PA(ci caelestia lura sacerdos)/COMISSOS PO(pulos moribus instituis.)/GLORIA VOS M(aior laudet coronat)/VIRTVTVM M(ater felicior subole). O'Donnell (1992c: 142) remarked that Monnica's remains were removed in 1430 from her burial place in the church of S. Agostino in Rome by Pope Martin V.

2. Although pointing out that *Confessions* is not a historical or biographical document, but a theological demonstration of God's intervention in the life of a sinner, Courcelle (1968: 40–43) lists the omissions and silences that could furnish a whole program of research on the historicity of Augustine's book. The interpreter would have a lot to choose from; questions demanding answers would range from the absence of concrete references to historical events and characters, like the Emperor to whom Augustine delivered a panegyric in Milan, to who was the dear friend, whose death caused him so much pain. The silences and omissions provide a wealth of intentions to be explained, if the goal is to identify the personal purpose and the meaning of *Confessions.* Throughout Augustine's autobiographical narrative, the historical and the lived reality of a specific individual were reframed under the theological and moral canon defining and forming the life of a Christian.
3. In *The City of God* (*civ* 5, 26) Augustine returned to the struggle between Maximus and Theodosius, and again made no reference to his feelings about the blockade. What mattered to Augustine was that Theodosius, since the beginning of his reign, favoured the Church with his endeavours.
4. While Chadwick translated "*cuiusdam Ciceronis*" for "a certain Cicero," Sheed preferred "one Cicero." Testard (1958: 11–19) compared this passage to many others in which Augustine lavishes praise on Cicero, albeit restricted to his mastery of rhetoric, and then concluded that the words could possibly be a device to call the attention of the reader. Cicero's fame, as a writer, however, made such verbal use futile and quite strange. O'Meara (1992: 90) saw Augustine's use of equivalent words under a different light. Speaking of Aeneas, Augustine wrote "*nescio quis*," "some Aeneas or other," when he certainly knew the mythological and literary character of Virgil's *Aeneid.*

Similar words are used in reference to Paul, the Apostle: Augustine employs "*quidam*," a certain person. The fig tree in the conversion scene in a garden in Milan and the window when the vision of Ostia occurs are also referred to in *Confessions* as *quaedam*, literally "a certain fig tree," "a certain window." O'Meara argues that Augustine's intention was to place "the emphasis and concentration off the nouns to which they were attached, in order to heighten the providential in the scene."

5. Brown (1967: 70) understood that "Symmachus had every reason to welcome a non-Catholic in so an important post." Barnes (1992) disagrees, identifying Symmachus and Ambrose, the Catholic Bishop of Milan, as probably first cousins, having exchanged respectful letters, and attended mutual requests, although "cool and distant" (Barnes 1992: 9). Symmachus and Ambrose represented opposing sides on the conflict of the Altar of Victory, in the church of Milan, when the Arian empress Justina, wife of Valentin I and mother of Valentin II, ordered the invasion of the church, resisted by Catholics, among them the devout mother of Augustine (*conf.* 9, vii, 15.) Barnes concludes that the connections with both Symmachus and Ambrose were useful to Augustine: for a provincial young man, "it was his conversion which brought Augustine rapid social mobility in a Christian society" (Barnes 1992: 11).
6. Courcelle's (1968) detailed and insightful analysis of Augustine's mystical quest and his vain attempts at Platonic ecstasy showed that the whole Augustinian method could be traced to Neo-Platonic books, translated into Latin by Marius Victorinus. The translations of Marius Victorinus may have come to Augustine's hands through Simplicianus and Ambrose. In *The City of God (civ.* 9, 16), Augustine openly subscribed to the Platonic notion that God is the sole being that cannot be defined by words, because human language is incapable of expressing His totality. The personal effort of introspection should be immense, but to reach God is feasible if we transcends our corporeal nature and apprehend Him in a sudden flash of dazzling light.
7. In his analysis of Augustine on signs, Markus (1952: 83) identifies symptom to index. Symptoms are *signa naturalia*.
8. It must be remembered at this point that a Christian and an ancient person such as Augustine would not consider allegories as debased or contrived fictions. As he stated so clearly in his commentaries to Genesis, meaning is both literal and figurative.

9. For a glimpse of the debate among Augustinian scholars about the reality of the voice in the narrative of *Confessions*, compare Courcelle's arguments in his *Recherches sur les Confessions de Saint Augustin* (1968: 188–202 and 291–310) with O'Donnell's (1992c: 59–65) more orthodox defense of the reality of Augustine's description.
10. Against the idea of Ambrose as the inventor of silent reading, expressed by Balogh (1926), see Knox (1968.)

Part 2
Ancestral games

4. Origin

1. Malinowski's claim that the fieldworker should take a respectful peek inside the studied culture did not prevent the founding father of anthropology from expressing extreme irritation and contempt for the natives. In his posthumously published diary, Malinowski (1967: 282) would refer to them as *niggers*: "In the morning worked for two hours at Teyvava; felt very poorly and very nervous, but didn't stop for a moment and worked calmly, ignoring the niggers." To be completely fair, we must mention Stocking (1983) about the difficulty in translating mechanically the original word in Malinowski's diary, *nigrami*, into the infamous *nigger*. *Nigrami* as a word composed of the English racial epithet, *nigr*, and *ami* from the Polish could be ambiguous and puzzling. Yet Stocking (1983: 102) believes that this is "no reason to argue that the word did not have a derogatory racial meaning." On the other hand, Leach (1980) defends Malinowski, claiming that the *Dairy*, as it was published, is an unreliable source of Malinowski's ideas. The *Diary* goes from March 1915 to March 1916, and at this time, Malinowski had not started his Trobriand research. That is a reasonable point; but to dismiss, as Leach (1980: 2) does, the translation of *nigrami* for nigger, arguing that *nigrami* "could not have carried, in 1918, the special loaded meaning which the term nigger conveys to American readers of 1970" is an exaggeration. Since the end of the seventeenth century and beginning of the eighteenth century, the term had a contemptuous connotation. The *Oxford English Dictionary* quotes a line from Lord Byron (1788–1824) with this meaning. The same condescending attitude as Malinowski's can be seen in Radcliffe-Brown's (1958) definition of anthropology as "of practical value in connection with the administration of backward people."

2. Leopold (1980) provides a detailed description of Tylor's intellectual development and background.
3. Franz Boas was another major force urging the anthropologist to avoid generalizations in favour of detailed studies of particular cultures. For an analysis of the impact of the Boasian approach, see Brown (1991: 54–58).
4. Sahlins (1976) depicts Malinowski's conception of language as the bastardization effect of a narrow pragmatic idea of meaning. Malinowski's insistence on immediate lived experience created a pernicious division in anthropological thought. As a result of this request, we would find conventions and rules that make up cultural dimensions existing in a different realm from the actual behaviour of social actors (Sahlins 1976: 80). For Sahlins's hard-core conventionalism, that division was an obstacle hampering both cultural theory and anthropology as a whole.
5. About half of the world's population is immersed in an Indo-European linguistic universe. Latin, Germanic, Celtic, and Slavic languages are part of the Indo-European linguistic tree, as well as Greek, Albanian, Armenian, Iranian, Gypsy, Baltic, and some of the languages spoken in India (Malherbe 1983: 134). Among other traits of the Indo-European languages, such as the fact that words with a fixed form (adverbs, prepositions) are less numerous than the ones that suffer some kind of flexion (nouns, pronouns, verbs), Malherbe (1983: 135) also indicates that the subject determines the conjugation of the verb; the complement plays no role in defining the verbal form. The subject is essential; the predicate will follow it.
6. Refusing to acknowledge that links between language and thought can be reduced to a mere set of superficial distinctions, such as the claim that thought is universal and language particular, Benveniste (1971: 55–64) examined the interaction of language and philosophy. In the Aristotelian system of categories, for instance, the term *ousia* means substance or essence, but it also applicable to linguistic names signifying a class of objects. Not only essences but also all other Aristotelian categories arise from language itself. Benveniste's point is clear: linguistic categories conduct cognitive assumptions, at least in the case of classical philosophy. The appropriateness of linguistic categories comes mainly from the familiarity of linguistic expression. Therefore, "no matter how much validity Aristotelian categories may have as categories of thought, they turn out to be transposed from categories of language" (Benveniste 1971: 61). Earlier, in *The Twilight of Idols* (1990:

48), Friedrich Nietzsche saw that categories of reason are projections of language. After noticing the influence of philosophical reasoning on theological assumptions, he observed sarcastically: "I fear we are not getting rid of God because we still believe in grammar."

7. Wickler (1968) discusses in detail the mechanics of mimetic attack and defense both in the animal world and among plants.
8. Williams's (1964) experiments on the consociation of fish conclude that group formation is an alternative even for fish that, in their original niches, do not exercise this kind of behaviour.

5. Sex, signals

1. Leach's all too brief remark about the subject – although symptomatic of an important objection – was made during an interview for a fascinating history of the sociobiology controversy: *The Defenders of the Truth* (Segerstråle). As the conversation unfolded, Leach observed that societies have generated countless combinations of kinship ties, mainly if the analyst takes into account cousin marriages. Female choice is just too narrow to explain the overwhelming diversity. Then, further attacking the idea of choice as investment in offspring, Leach finds curious the idea "that by and large individuals can somewhat *choose* their mates! In most of the world they don't! Their love affairs are different from their marriages. The marriages are arranged for political reasons" (Segerstråle 2000: 173). Knowing that Leach defined himself as a structuralist in anthropology, I have developed in the language of structural anthropology what he might have said.
2. Miller (2000: 45–67) presents the steps that the idea of sexual selection had to struggle to be accepted as a topic worthy of interest. Miller makes an excellent case about what the human sciences lost after disregarding sexual selection.
3. The individual female cannot do much to prevent the additional costs of having males. Nonetheless, this question is justified from an evolutionary perspective, which means considering long stretches of time. During the unfolding of life processes, it is hard to believe that no other mutated alternative could have appeared to benefit females.
4. The genius of Darwin is that he saw with great acumen the importance of the sex ratio as a major biological question to be tackled. The explanation in this chapter blends the original work of Ronald A. Fisher (1958), George C. Williams (1975, 1992), John Maynard Smith (1971 and 1978), John Maynard Smith and Eörs Szathmáry (1995: 147–167

and 1999: 78–93,) and W. D. Hamilton (2001). An important contribution to the inescapably competitive nature of sex is the *Red Queen hypothesis,* a direct reference to Lewis Carroll's *Through the Looking Glass.* Sex is paradoxically competitive, in a way similar to Carroll's notion that in the *Looking Glass* world, to remain in place one should run as much as possible, for the environment and the other people are doing the same. To outpace the others the effort should even greater. Mutual and restless competition is therefore at the core of an evolving world. Leigh van Valen (1973) was the first to recognize this phenomenon as an evolutionary principle. See Ridley (1994) for a comprehensive review of the *Red Queen hypothesis.*

5. Zahavi and Zahavi (1997) contend forcefully that mating and displaying the intention of fighting for a territory make sense only if they are conveyed through honest and reliable signals. The animals involved in both mating and fighting will interpret false signals of biological excellence, and in the end, such displays would be ineffectual, and therefore would not spread in a population. Also, if the animal is weak, to present willingness for conflict can only be disastrous.
6. Jones (1983) as well as Powell and Jones (1983) contend that the case of attraction in flower mimicry is rather different from what happens in warnings, repulsion, or camouflage between predators and prey. In floral mimicry, the cost of attraction is not the life of the attracted animal, but simply not receiving the rewards promised by the signal. The conflict of interest does not lead to annihilation. Food is not gathered, and yet the mimic is eventually pollinated. Jones (1983) argues that the following conditions are present in floral mutualism: food is present in all organisms in all *taxa*; all *taxa* are apparently models, which means that deception is not necessarily involved; the pollinators are unspecialized and apparently not constant to any given species. Floral mutualism does not exclude, but its mechanism is akin to optimal advertising. Jones (1983: 296) observes, "[D]eceit and mutualistic mimicries are part of a continuum." Again, continuity is the dominant mechanism of evolution itself.
7. In the 1970s, when the concept of the handicap principle was first formulated, the reaction was disbelief and rejection. Referring to the Zahavis' theory in the first edition of *The Selfish Gene* (1976), Richard Dawkins argued that a handicap could not be an adaptive trait capable of spreading itself among organisms of a species. A real handicap must be a biological penalty, and therefore it has to be undesirable. Another reason for the initial dismissal of the handicap principle was that the

Zahavis' formulation was strictly verbal, and did not present a mathematical model to justify it. The lack of a mathematical model in the Zahavis' papers on the handicap principle was an indication of its limited capacity of prediction. Everything sounded suspiciously anecdotal. Mentioning his own problems with the handicap principle, W. D. Hamilton (2001: 210–211) also hammers on the lack of a mathematical model, and presents other objections, one of them terminological: "a broken leg is clearly a handicap; is that what Zahavi wanted to say?" although Hamilton was ready to agree with the idea that costliness is a proof of honesty. After the Zahavis's polemic articles, Johnstone and Grafen (1993) demonstrated that the handicap principle could be represented mathematically, and then the objections among biologists to the concept were gradually reduced. In the second edition of *The Selfish Gene* (1989), Dawkins retracted from the original critique and admitted that the handicap principle was an acceptable theory, indeed a plausible one. Lately, Maynard Smith and Harper (2003) have insisted the notion of a handicap signal is totally dependent upon the cost involved in bearing it. Three basic assumptions interact: the female must be capable of inferring a male quality from the signal displayed; the signals must have a cost; and the costs are variable, thus implying that some males cannot pay its price. Being able to pay the high cost of a signal indicates a male of high biological quality.

Part 3
Individual games

6. Strategies

1. Poundstone's (1992: 51) excellent introduction to game theory uses the same example to illustrate utility.
2. Not long ago, in class, I realized that a graduate student was frowning at the explanation of matching pennies. At first, I did not understand why he was squinting so much. Later, he told me that he had misheard the name of the game and thought that it was *matching panties*, so nothing made sense. It would be, however, great fun to adapt the original child's game to this more risqué possibility. The players leave the room and change underwear. They undress in public and if the panties match, player *A* wins; if not, the opponent is the winner. For those with wilder imagination, I will leave the determination of what the prize will be.

7. Players

1. Recent advances in game theory have come to accept that the predictive success of a model does not have to depend on the identification of a single point of bargaining. Aumann and Maschler (1964) demonstrated that bargaining could occur – as it does so frequently – within regions of choices. To an analysis with experimental data, see Selten and Krishker (1982).
2. Robert Wright (2000) recognized that the passage from zero-sum to nonzero definitely shapes human destiny. Beginning with historical considerations, Wright extends the insight to the organic world.
3. Sen (1969, 1997: 74–83) advances an interesting solution to the quandaries of the Prisoner's Dilemma. He presumes that selfish players could cooperate if embracing a mode of Kantian categorical imperative. They could act in a mutually concerted manner and thus feel simultaneously the obligation to cooperate if they took their actions as guided by a universal principle. Shared universal moral principles would provide the necessary assurance so they could act in harmony without betraying their own interests. Sen insists that no enforcement would be needed in this case. The solution is ingenious, although odd if applied to the tale of criminals about to choose between cooperation and defection. It could well be that the exemplary tale of prisoners loads the reply and excludes immediately the possibility of moral cooperation. However, how would it be possible to account for the emergence of cooperation among organisms that do not have morality at all? They play Prisoner's Dilemma and they are part of cooperative wholes. Selfish zero-sum actors end up in nonzero relations. But how can that be? To reply to that question, it is quite acceptable to transfer Flood's story about uncooperative secretaries to the Serengeti plains. The laggard male lions are incorporated into the pride anyway despite their uncooperative tendencies: the lionesses are resigned and thus not different from Secretary 1 in Flood's experiment. Amartya Sen's solution to the Prisoner's Dilemma is mainly a vigorous critique of the mechanical transference of the individualistic setting of the game to the collective allocation of resources. The Prisoner's Dilemma should never be invoked to disqualify welfare programs. The Prisoner's Dilemma does not provide any criterion to judge welfare. From Sen's perspective – and undeniably so – welfare is a moral and social duty.
4. Two important exceptions are a paper by Riechert and Hammerstein (1983) that uses the competition for water in desert plants to analyze

the game and Nowak and Sigmund's (2004) outstanding interpretation of the evolutionary dynamics of biological games. Nowak and Sigmund (2004: 795) redefine the Prisoner's Dilemma and the action of an uncooperative player taking a free ride on some other participant of the game as "hawk-dove or chicken game."

5. Eumelus was an Archean, son of Admetus and Alcestis – commander of the Thessalians; Diomedes, the king of Argos; Antilochus, a Pylean prince, son of Nestor, and Menelaus, the king of the Lacedaemons, husband of Helen, the woman who was, according to the mythological sources, the cause of the Trojan War. The role of Achilles – the quintessential warrior – in Patroclus' funeral games is the one of agônothetês: "the sponsor, the president, the chief umpire of an agôn, or 'contest'" (Spivey 2004: 6).

Afterword

1. Lovisolo (2002) refers to the same prejudices in cultural interpretations, calling it deep socialization. Much earlier, Wrong (1961) presented a critique of this notion of extreme socialization.
2. Greek social distinctions conferred on individuals who excelled in warfare were a heavily sexualized notion of military valor as Cartledge (2003: 129) points out. Greek language had a definite word, *aristeia*, for the sort of military excellence that was displayed conspicuously in the battlefield and "was applied most famously to the deeds of Greek heroes related in Homer's *Iliad*, hence the *aristeia* of Diomedes, of Patroclus, and, above others, of Achilles." Moreover, in this sense, *aristeia* is a feminine singular noun, although "the Greeks used the same letters for a neuter plural meaning, not the valor itself, but the prizes for the valor, which they awarded competitively after battles such as those of Thermopylae and Plataea."
3. Aristophanes' choice of the name Lampito for the Spartan woman who, with Lysistrata, will begin a Pan-Hellenic alliance to end the war through sexual deprivation is especially meaningful. The character Lampito bears the same name as Archidamus' wife, daughter of Leotychidas (Cartledge 2003: 158). Another interesting touch in *Lysistrata* is how Athenian women recognize and admire the beauty of Lampito. She is sleek, slender, strong, with soft hands, and firm, lovely breasts, probably a clear reference to the fact that the fitness of Spartan women was the result of their freedom from the grind of domestic chores, including breastfeeding, which were left for the enslaved helots. More-

over, despite her astonishing looks, Lampito's tongue-tied manner of speaking reduces her sexual attractiveness.

References

Ackrill, J. L.
2000 Language and reality in Plato's *Cratylus*. In *Plato*, Gail Fine (ed.), 127–141. New York: Oxford University Press.

Adcock, Frank and Moseley, D. J.
1975 *Diplomacy in Ancient Greece*. New York: Saint Marin's Press.

Akerloff, George A.
1970 The market for "lemons": Quality uncertainty and the market mechanism. *The Quarterly Journal of Economics* 84, 488–500.

Aristophanes
1971 *The Complete Plays*. New York: Bantam Books. Translations from the Greek, edited with an introduction by Moses Hadas of works probably written between 427 and 388 B.C.E.

Aristotle
1967 *The Athenian Constitution*. In *Aristotle in 23 volumes. Volume XX*. Cambridge: Harvard University Press. Translation from the Greek by H. Rackham of a work written probably between 328 and 325 B.C.E. Abbreviation: *Ath. Const.*

1991 *On Rhetoric: A Theory of Civic Discourse*. New York: Oxford University Press. Translation from the Greek with introduction notes and appendices by George A. Kennedy of a work written between 360 and 334 B.C.E.

Arrow, Kenneth J.
1983 A difficulty in social welfare. In *Collected Papers of Kenneth J. Arrow, volume 1: Social Choice and Justice*, 1–29. Cambridge: The Belknap Press of Harvard University Press.

Augustine, Saint, Bishop of Hippo
1938 *Concerning the Teacher (De Magistro)*. New York: Appleton. Translation from the Latin by George C. Lecki of a work

written around 388–389 CE. The abbreviation of the title is *mag.*

1947a *The Advantage of Believing (De Utilitate Credendi).* In *Writings of Augustine, volume 2.* New York: CIMA. Translation from Latin by Luanne Meagher of a work written in 391 CE. Abbreviation: *util. cred.*

1947b *The Magnitude of the Soul (De Animae Quantitate).* In *Writings of Augustine, volume 2.* New York: CIMA. Translation from Latin by John McMahon of a work written probably in 388 CE. Abbreviation: *quant. an.*

1947c *On Faith in Things Unseen (De Fide Rerum Invisibilium).* In *Writings of Augustine, volume 2.* New York: Cima. Translation from Latin by Joseph Deferrari and Mary Francis McDonald of a work written around 400 CE. Abbreviation: *f. invis.*

1947d *On Music (De Musica.)* In *Writings of Augustine, volume 2.* New York: CIMA. Translation from Latin by Robert Catesby Taliaferro of a work written around 387 CE. Abbreviation: *mus.*

1947e *The Immortality of the Soul (De Immortalitate Animae.)* In *Writings of Augustine, volume 2.* New York:CIMA. Translation from Latin by Ludwig Schopp of a work written around 386-387 CE. Abbreviation: *imm. an.*

1947f *Enchiridion: Faith, Hope, and Charity (Enchiridion: Liber de Fide, Spe et Caritate.)* In *Writings of Augustine, volume 4.* New York CIMA. Translation from Latin by Bernard M. Pebles of a work written around 420–421 CE. Abbreviation: *ench.*

1948a *The Divine Providence and the Problem of Evil (De Ordine).* New York: CIMA. Translation from Latin by Robert P. Russell of a work written around 386–387 CE. Abbreviation: *ord.*

1948b *The Spirit and the Letter (De Spiritu et Littera).* In *Basic Writings of Saint Augustine, volume 1.* New York: Random

House. Translation from Latin and edited with introduction, and notes by Whitney J. Oates of a work written in 412 CE. Abbreviation: *spir. et litt.*

1948c *The Happy Life (De Beata Vita).* New York: CIMA. Translation from Latin by Ludwig Schopp of a work written between 386 and 387 CE. Abreviation: *beata. v.*

1950 *Against the Academics (Contra Academicos).* Westminster: The Newmann Press. Translation from Latin with notes by John J. O'Meara of a work written in 386–387 CE. Abbreviation: *c. acad.*

1952 *Lying (De Mendacio).* In *Treatises on Various Subjects*, Robert Deferrari (ed.). New York: Fathers of the Church. Translation from Latin by Sister Sarah Muldowney of a work written in 395 CE. Abbreviation: *mend.*

1953 *Letters (Epistulae).* In *Letters, volume 2 (81–130); and volume 3 (131–164.)* New York: Fathers of the Church. Translated from Latin by Sister Wilfrid Parsons. Abbreviation: *ep.*, followed by the number of the letter.

1955a *The Care to be Taken for the Dead (De Cura Pro Mortuis Gerenda).* In *Treatises on Marriage and other Subjects.* NewYork: Fathers of the Church. Translation from the Latin by John A. Lacy, edited by Roy Deferrari of a work written around 422 CE.

1955b *Faith and the Creed (De Fide et Symbolo).* In *Treatises on Marriage and other Subjects.* New New York: Fathers of the Church. Translation from the Latin by Robert R. Rusell, edited by Roy Deferrari of a work in 393 C.E. Abbreviation: *f. et sym.*

1957 *Against Julian (Contra Julianum).* New York: Fathers of the Church. Translated from Latin by Matthew A. Schumacher of a work written around 431 CE. Abbreviation: *c. Iul.*

1962 *The Trintiy (De Trinitate.)* Washington DC: The Catholic University Press of America. Translation from Latin by

Stephen McKenna of a work began in 401 and finished in 416 CE. Abbreviation: *trin.*

1967 *Concerning the City of God Against the Pagans (De Civitate Dei).* Harmondsworth: Penguin. Translation from Latin by Henry Bettenson of a work written at intervals between 413 and 427. Abbreviation: *civ.*

1968 *The Retractions (De Retractiones.)* Washington DC: The Catholic University of America Press. Translation from Latin by Mary Inez Bogan a work written in 426 or 427. Abbreviation: *ret.*

1977 *On Chrstian Doctrine (De Doctrina Christiana).* New Jersey: Prentice Hall. Translation from Latin with an introduction by D. W. Robertson of a work began in the mid-390s, interrupted and finished in the late 490s. Abbreviation: *doctr. chr.*

1982a *The Literal Meaning of Genesis* (*De Genesi ad Litteram Libri*), *volume 1, Books 1–6.* New York: Newman Press. Translation from the Latin by John Hammond Taylor of a work began in 399 and finished by 415.C E.. Abbreviation: *Gn. Litt.*

1982b *The Literal Meaning of Genesis* (*De Genesi ad Litteram Libri*), *volume 2, Books 7–12.* New York: Newman Press. Translation form the Latin by John Hammond Taylor of a work began in 399 and finished in 415 C.E. Abbreviation: *Gn. litt.*

1991 *Confessions (Confessiones).* New York: Oxford University Press. Translation from Latin with an introduction by Henry Chadwick of a work written probably written between 397 and 401. Abbreviation: *conf.*

1993 *Confessions; Books I-XIII (Confessiones).* Indianapolis: Hackett. Translation from Latin by F. S. Sheed, with an introduction by Peter Brown of a work probably between 397 and 401. Abbreviation: *conf.*

1997 *On Christian Instruction (De Doctrina Christiana).* Oxford: Clarendon Press. Translation from Latin and edited by R. P. H.

Green of a work began in the mid 390s, interrupted, and finished in the late 420s Abbreviation: *doctr. chr.*

2000 *Soliloquies: Augustine's Interior Dialogue (Sololiquia)*. New York: New York City Press. Translation from Latin with notes by Kim Paffenroith, with an introduction by Boniface Ramsey, edited by John Rotelle. Abbreviation: *sol.*

Aumann, Robert J.

2000 *Collected Papers.* Cambridge: The M.I.T. Press.

Aumann, R. J., and Maschler, M.

1964 The set bargaining for cooperative games. In *Advances in Game Theory*, M. Drescher, L. S. Shapley, and A. W. Tucker (eds), 443–476. Princeton: Princeton University Press.

Axelrod, Robert

1984 *The Evolution of Cooperation*. New York: Basic Books.

Axelrod, Robert and Hamilton, W. D.

1981 The evolution of co-operation. *Science* 211, 1390-1386. Reprinted in Axelrod (1984) and Hamilton (2001).

Balogh, Josef

1926 "Voces paginarum": Beiträge sur Geschichte des Lauten Lesen und Schreibens. *Philologus* 82, 84–109.

Balter, Michael

2005 Are human brains still evolving? Brain genes show signs of selection. *Science* 309, 1662–1663.

Barnes, T. D.

1992 Augustine, Symmachus, and Ambrose. In *Augustine: From Rhetor to Theologian*, Joanne McWilliam, Timothy Barnes, Michael Fahey, and Peter Slater (eds.), 7–15. Waterloo: Wilfrid Laurier Press.

Barthes, Roland

1964 *Elements de Sémiologie.* Paris: Éditions du Seuil.

1995 *Leçon*. In *Oeuvres Complétes, tome III: 1974–1980,* Éric Marty (ed.), 801–814. Paris: Éditions du Seuil. Text of the inaugural class at the Collége de France, 1978.

Benveniste, Émile
1966 Problèmes de Linguistique Générale. Paris: Éditions Gallimard.

Binmore, Ken
1990 *Essays in the Foundations of Game Theory.* Cambridge: Blackwell.

1994 *Game Theory and the Social Contract, volume 1: Playing Fair.* Cambridge, Ms: The MIT Press.

Bonner, John Tyler
1980 *The Evolution of Culture in Animals.* Princeton: Princeton University Press.

Bosworth, A. B.
2000 The historical context of Thucydides's Funeral Oration. *Journal of Hellenic Studies* 10, 1–16.

Bouissac, Paul
2004a Does Saussure still matter? *The Semiotic Review of Books* 14 (3), 10–12.

2004b Saussure's legacy in semiotics. In *The Cambridge Companion to Saussure,* Carol Sanders (ed.), 240–260. Cambridge: Cambridge University Press.

Boyd, Robert., and Loberbaum, Jeffrey P.
1987 No pure strategy is evolutionarily stable in the repeated Prisoner's Dilemma. *Nature* 327, 58–59.

Bradbury, J. W., and Veherencamp, Sandra
1998 *Principles of Animal Communication.* Sunderland, MA: Sinaeur.

Brandwood, Leonard
1990 Stylometry and chronology. In *The Cambridge Companion to Plato*, Robert Kraut (ed.). Cambridge: Cambridge Uiversity Press.

Bronowski, Jacob
1973 *The Ascent of Man*. Boston: Little, Brown, and Company.

Brown, Donald E.
1991 *Human Universals*. New York: McGraw-Hill.

Brown, Peter
1967 *Augustine of Hippo: A Biography*. Berkeley: The University of California Press.

1993 Introduction. In *Augustine: Confessions; Books I-XIII.* Indianapolis: Hackett.

Brunt, P. A.
1993 *Studies in Greek History and Thought.* Oxford: Clarendon Press.

Cartledge, Paul
2002 *The Spartans: The World of the Warrior-Heroes of Ancient Greece.* Woodstock, N. Y.: Overlook Press.

Casamassa, Antonio
1952 Ritrovamento di parte dell'elogio di S. Monica. *Atti della Pontificia Accademia di Archeologia, Rendiconti* 152–154, 271–273.

Cheney, Dorothy L. and Seyfarth, Robert M.
1991 Truth and deception in animal communication. In *Cognitive Ethology: The Mind of other Animals,* Carolyn A. Ristau (ed.), 127–151. Hillsdale, NJ: Lawrence Erlbaum.

Cicero, Marcus Tullius
2001 *On the Ideal Orator (De Oratore).* New York: Oxford University Press.Translated from Latin with introduction,

notes, appendixes, glossary, and indexes by James M. May and Jakob Wisse of a work written circa 55 B.C.E.

Cogan, Marc
1981 *The Human Thing: The Speeches and Principles of Thucydides.* Chicago: The University of Chicago Press.

Connor, W. R.
1989 Early Greek warfare as symbolic expression. *Past and Present* 119, 3–29.

Courcelle, Pierre
1968 *Recherches sur les Confessions de Saint Augustine.* Paris: Boccard.

Cournot, A. A.
1960 *Researches into the Mathematical Principles of the Theory of Wealth.* New York: A. M. Kelley. Translation from the French by Nathaniel T. Bacon of a work originally published in 1838.

Darwin, Charles
1979 Reprint. *On the Origin of Species.* New York: Gramercy. Original Edition, London: J. Murray, 1859.

1987 *Charles Darwins' Notebooks, 1836 - 1844. Geology, Transmutation of Species Metaphysical Enquiries*, P. H. Barret, P. J. Gautrey, Herbert D. Kohn, and S. Smith (eds.) Ithaca and New York: British Museum (Natural History) and Cornell University Press.

1998a Reprint. *The Expression of Emotions in Man and Animals, third edition;* introduction, afterword, and commentary by Paul Ekman. New York: Oxford University Press. Original edition, London: John Murray, 1872.

1998b Reprint. *The Descent of Man; and Selection in Relation to Sex.* Introduction by H. James Brix. NewYork: Prometheus Books. Original edition, New York: Crowell.

Dawkins, Richard

1983 Replicators and vehicles. In *Current Problems in Sociobiology,* King's College Sociobiology Group (ed.), 45–64. Cambridge: Cambridge University Press.

1989 New edition. *The Selfish Gene.* New York: Oxford University Press. Original edition, Oxford: Oxford University Press, 1976.

1996 *Climbing Mount Improbable.*New York: Norton.

Dihle, Albrecht

1982 *The Theory of Will in Classical Antiquity.* Berkeley: University of California Press.

Dorowszewski, Witold

1933 Quelques remarques sur les rapports de la sociologie et de la linguistique: Durkheim et Saussure. *Journal de Psychologie Normal et Pathologique* 30, 82–91.

Dumont, Louis

1967 Caste: A phenomenon of social structure or an aspect of Indian Culture. In *Caste and Race: Comparative Approaches.* Anthony de Rueck and Julie Knight (eds.), 28–38. Boston: Little Brown and Company.

1970 *Homo Hierarchicus: An Essay on the Caste System.* Chicago: The University of Chicago Press. Translated from the French by Mark Sainsbury. Original edition, Paris: Éditions Gallimard, 1966.

1983 *Essais sur l'Individualisme: Une Perspective Anthropologique sur l'Idéologie Moderne.* Paris: Éditions du Seuil.

Durkheim, Émile

1960 *Les Règles de la Méthode Sociologique.* Paris: Alcuin. Original edition published in 1895.

Engels, J.

1962 La doctrine du signe chez Augustine. *Studia Patristica* 6, 366–373.

Evans, Patrick D., Gilbert, Sandra I., Nitzan, Mekel-Bobrov, Vallender, Eric J., Anderson, Jeffery R., Vaez-Azizi, Leila M., Tishkoff, Sarah A., Hudson, Richard R., and Lahn, Bruce T.

2005 *Microcephalin*, a gene regulating brain size continues to evolve adaptively in Humans. *Science* 309, 1717, 1720.

Finley Jr., John H.

1963 *Thucydides*. Ann Arbor: The University of Michigan Press.

1967 *Three Essays on Thucydides*. Cambridge: Harvard University Press.

Fisher, Ronald A.

1958 Reprint. *The Genetical Theory of Natural Selection*. New York: Dover. Original edition, Oxford: Oxford University Press, 1930.

Flood, Merrill M.

1958 Some experimental games. *Management Science* 5(1): 5–26.

Fortin, Ernest L.

1975 Augustine and the problem of Christian rhetoric. *Augustinian Studies* 5, 85–100.

Foucault, Michel

1972 *The Archeology of Knowledge*. London: Tavistock. Translation from the French by A. Sheridan-Smith. Original edition, Paris: Éditions Gallimard, 1969

Gass, William H.

1989 *Habitations of the Word*. New York: Simon and Schuster.

Geertz, Clifford

1973 *The Interpretation of Cultures*. New York: Basic Books.

Gehrke, Hans-Joachim
1996 La stasis. In *I Greci: Storia, Cultura, Arte, Societá, v. 2: Una Storia Greca, Il Definizione*, Salvatore Setti (ed.), 453–480. Torino: Einaudi.

Ghiselin, Michael T.
1974 *The Economy of Nature and the Evolution of Sex.* Berkeley: The University of California Press.

Gilson, Etienne
1943 *Introduction a l'Étude de Saint Augustine.* Paris: J. Vrin.

1960 The Christian Philosophy of Saint Augustine. New York: Random House. Translation of Gilson (1943) from the French by L. E. M. Lynch

Ginzburg, Carlo
1989 *Clues, Myths, and the Historical Method.* Baltimore: Johns Hopkins University Press. Translation from the Italian by John and Anne Tedeschi. Original edition, Torino: Giulio Einaudi. 1986.

Godel, Robert
1957 *Les Sources Manuscriptes du Cours de Linguistique Génerale de F. de Saussure.* Paris: Minard.

Gomme, A. W.
1956a *A Historical Commentary on Thucydides, volume 1: Introduction and Commentary on Book I.* Oxford: Oxford University Press.

1956b *A Historical Commentary on Thucydides. Volume 2: Books I-III.* Oxford: Oxford University Press.

Griffin, Jasper
2003 It is all Greek! *The New York Review of Books* 1(20), 62–65.

Gwynn, Aubrey
1966 Reprint. *Roman Education from Cicero to Quintillian.* New York: Teacher's College Press. Original edition, 1922.

Haig, David
1993 Genetic conflicts in human pregnancy. *The Quarterly Review of Biology* 68: 495–532.

Hamilton, W. D.
1980 Sex versus non-sex versus parasites. *Oikos* 35, 282–290. Reprinted In Hamilton (2000).

1996 *Narrow Roads of Gene Land, volume 1: Evolution of Social Behavior.* Oxford: W. H. Freeman/Spektrum.

2001 *Narrow Roads of Gene Land, volume 2: The Evolution of Sex.* Oxford Oxford University Press.

Hamilton, W. D., Axelrod, Robert, and Tanese, Keiko
1990 Sexual reproduction as an adaptation to resist parasites (a review). *Proceedings of the National Academy of Sciences, USA,* 87, 3566–3573.

Hanson, Victor Davis
1989 *The Western Way of War: Infantry Battle in Classical Greece.* New York: Alfred Knopf.

1991 *Culture and Carnage: Landmark Battles in the Rise of Western Power.*New York: Doubleday.

1998 *War and Agriculture in Classical Greece.* Berkeley: The University of California Press.

Harris, Roy
1999 Saussure for all seasons. *Semiotica* 131 (3–4), 273–287.

2000 *Saussure and his Interpreters.* Edinburgh: Edinburgh University Press.

Harsanyi, John C.
1989 Bargaining. In *The New Palgrave: Game Theory,* John Eatwell, Murray Milgate, and Peter Newman (eds.), 51–66. London: W. M. Norton.

Hasson, O.
1992 Cheating signals. *Journal of Theoretical Biology* 167, 223–238.

Hegel, G. W. F.
1949 Revised second edition. *The Phenomenology of Mind.* London: George Allen snd Unwin. Translated from the German with an introduction and notes by J. B. Baillie. Original edition published in 1807.

Heidegger, Martin
1996 *Being and Time.* Albany, NY: State University of New York Press. Translated from the German by John Staumbaugh. Original edition, Tubingen, Niemeyer, 1927.

Henry, Pierre
1938 *La Vision D'Ostie.* Paris: Vrin.

1981 *The Path to Transcendence: from Philosophy to Mysticism in Saint Augustine.* Pittsburgh, PA: Pickwick Press. Translation from the French of Henry (1938) by Frances F. Burch.

Herodotus
1954 *The Histories.* Harmondsworth: Penguin. Translation from the Greek by Aubrey Sélincourt from a work written probably in the beginning of the century B.C.E.

Hirsh Jr., E. D.
1982 Beyond conventions. *New Literary History* 14, 389–397.

Hobbes, Thomas
1951 *The Leviathan.* Oxford: Oxford University Press. Original edition, 1651.

Holte, Rangan
1962 *Béatitude et Sagesse: Saint Augustine et le Probléme de la Fin de l'Homme dans la Philosophie Ancienne.* Paris: Études Augustinnienes.

Homer
1990 *The Iliad.* Translation from the Greek by Robert Fagles, with an introduction and notes by Bernard Knox of a work probably written at the end of the eighth century B.C.E.

Hornblower, Simon
1987 *Thucydides.* Baltimore: Johns Hopkins Univeristy Press.

1991 *A Commnentary on Thucydides, volume 1: Books I-III.* Oxford: Clarendon Press.

Hull, David. L.
1989 *The Metaphysics of Evolution.* Albany, NY: State University Press of New York.

Hume, David
1969 Reprint and new edition. *A Treatise of Human Nature,* edited with an introduction by Ernest C. Mossner. Harmondsworth: Penguin. First edition published in 1739 and 1740.

Huntington, Samuel P.
1993 The clash of civilizations? *Foreign Affairs* 72 (3), 22–49.

1996 *The Clash of Civilizations and the Remaking of World Order.* New York: Simon and Schuster.

Jaeger, Werner
1939 *Paideia: The Ideals of Greek Culture, volume 1: Archaic Greece, the Mind of Athens. New York: Oxford University Press.* Translation from the German second edition by Gilbert Highet.

1943 *Paideia: The Ideals of Greek Culture, volume 2: In Search of the Divine Center.* New York: Oxford University Press. Translation from the German manuscript by Gilbert Highet.

Jakobson, Roman
1991 *On Language,* Linda R. Waugh and Monique Monville-Burstin (eds.) Cambridge: Harvard University Press.

Johnstone, R. A., and Grafen, A.
1993 Dishonesty and the handicap principle. *Animal Behavior* 47, 759–764.

Jones, R. J.
1983 A review of floral food deception mimicries with comments on floral mutualism. In *Handbook of Experimental Pollination Biology*, C. E. Jones and R. J. Little (eds.), 294–309. New York: Scientific and Academic Editions.

Kagan, Donald
1969 *The Outbreak of the Peloponnesian War*. Ithaca NY: Cornell University Press

1974 *The Archidamian War*. Ithaca, NY: Cornell University Press.

1991 *Pericles of Athens and the Birth of Democracy*. New York: Free Press.

2003 *The Peloponnesian War*. New York: Viking.

Kahn, Herman
1960 *On Escalation: Metaphors and Scenarios*. New York: Praeger.

Kavane, Eugene
1967 Augustine's *De Doctrina Christiana*: A treatise of Christian education. *Recherches Augustiniennes* 6, 97–133.

Ketchum, Richard J.
1979 Names, forms, conventionalism: *Cratylus* 383–395. *Phronesis* 24, 133–147

Khare, R. S.
1971 Encompassing and encompassed: A deductive theory of caste system. *The Journal of Asian Studies* 30: 859–868.

Knox, Bernard M. W.
1968 Silent reading in Antiquity. *Greek, Roman, and Byzantine Studies* 9, 421–435.

Koerner, E. F. K.
1973 *Ferdinand de Saussure: Origin and Development of his Lingustic Thought in Western Studies of Language*. Berlin: Vieweg.

Kraut, Richard
1991 Introduction to the study of Plato. In The Cambridge Companion to Plato, Richard Kraut (ed.), 1–50. Cambridge: Cambridge University Press.

Krebs, J. R. and Dawkins, Richard
1983 Animal signals: Mind-reading and manipulation. In *Behavorial Ecology:An Evolutionary Approach*, J. R. Krebs and N. B. Davies (eds.), 380–402. Sunderland, MA: Sinauer.

Krentz, Peter
1997 The strategic culture of Periclean Athens. In *Polis and Polemos,* Charles Hamilton and Peter Krentz (eds.), 55–72. Claremont: Regina Books.

Kuhn, Harold W. and Nassar, Sylvia (eds.)
2002 *The Essential John Nash*. Princeton: Princeton University Press.

Kuhn, Thomas S.
1965 *The Structure of Scientific Revolutions.* Chicago: The University of Chicago Press.

Lacan, Jacques
1976 *Écrits: A Selection.* New York: W. W. Norton. Translation from the French by Alan Sheridan. Original edition, Paris: Éditions du Seuil, 1966.

Laland, Kevin N. and Hoppit, William
2003 Do animals have culture? *Evolutionary Anthropology* 12, 150–159.

Leach, Edmund R.
1957 The epistemological background of Malinowski's empiricism. In *Man and Culture: An Evaluation of the Work of Bronislaw*

Malinowski, Raymond Firth (ed.), 119–138. London: Routledge and Kegan Paul.

1980 On reading *A Diary in the Strict Sense of the Term*; or the self-mutilation of Professor Hsu. *Rain* 36:2–3.

1984 Glimpses of the unmentionable in the history of British anthropology. *Annual Review of Anthropology* 13: 1–23.

Leopold, Joan
1980 *Culture in Comparative and Evolutionary Perspective: E. B. Tylor and the Making of Primitive Culture.* Berlin: Dietrich Reimer Verlag.

Lévi-Strauss, Claude
1945 L'analyse struturelle en lingustique et anthropologie. *Word. Journal of the Linguistic Circle of New York* 1(1), 1–21. Reprinted in Lévi-Strauss (1966).

1946 La technique du bonheur. *Esprit* 127: 643–657.

1962a *Totemisme, Aujourd'hui.* Paris: Presses Universitaires de France.

1962b *La Pensée Sauvage.* Paris: Librarie Plon.

1966 *Structural Anthropology.* New York: Basic Books. Translation from the French by Claire Jakobson and B. Grundfest-Schoepf. Original edition, Paris: Librairie Plon, 1958.

1976 *Structural Anthropology, volume 2.* New York: Basic Books. Translated from the French by Monique Layton.. Original edition, Paris: Librairie Plon, 1973.

1988 *The View From Afar.* New York: Basic Books. Translated from the French by Joachim Neugroschel and Phoebe Hoss. Original edition, Paris: Librarie Plon, 1983.

Lewis, David K.
1970 *Convention: A Philosophical Study*. Cambridge: Harvard University Press.

Lewontin, R. C.
1983 The organism as the subject and the object of evolution. *Scientia* 118, 65–82.

Loreaux, Nicole
1986 *The Invention of Athens: The Funeral Oration in the Classical City*. Cambridge: Harvard University Press. Translation from the French by Alan Sheridan. Original edition, Paris: Éditions de l'École des Hautes Études en Sciences Sociales, 1981.

Louth, Andrew
1981 *The Origins of the Christian Mystical Tradition: From Plato to Denys*. Oxford: Clarendon Press.

Lovisolo, Hugo
2002 Um homem chamado cavalo: Notas sobre a socialização [A man called horse: Notes on socialization]. In *A Sociedade na Tela do Cinema: Imagem e Comunicação* [Society on the Movie Screen : Image and Communication], Heris Arnt and Ronaldo Helal (eds.), 9–18. Rio de Janeiro: E-Papers.

Luce, J. V.
1964 The theory of ideas in the *Cratylus*. *Phronesis* 10, 21–36.

MacKenzie, Margaret
1983 Putting the *Cratylus* in its place. *Classical Quarterly* 36 (1), 124–150.

McComskey, Bruce
2002 *Gorgias and the New Sophistic Rhetoric*. Carbondale: Southern Illinois University Press.

Madec, Goulven
1975 Analyse du *De Magistro*. Revue des *Études Augustinnienes* 21, 63–71.

Malherbe, Michel
1984 *Les Languages de l'Humanité.*Paris: Seghers.

Malinowski, Bronislaw
1922 *Argonauts of the Western Pacific: An Account of Native Enterprise and Adventure in the Archipelagoes of Melanesian New Guinea.* Loodon; Routledge and Kegan Paul.

1923 The problem of meaning in primitive languages. In *The Meaning of Meaning,* C. K. Ogden and I. A. Richards (eds.), 296–336. New York: Harcourt, Brace and Company.

1931 Culture. In *Encyclopedia of the Social Sciences, volume 6,* 621–646. New York: Macmillan.

1967 *A Diary in the Strict Sens of the Term.* New York: Harcourt, Brace and World. Malthus, Thomas R.

Malthus, Thomas R.
1888 *An Essay on the Principle of Population; or a View of its Past and Present Effects on Human Happiness.* London: Reeves and Turner.

Markus, R. A.
1952 Saint Augustine on signs. *Phronesis* 2, 60–83.

Marrou, H. I.
1936 *Saint Augustine et la Fin de la Culture Antique* Paris: Boccard.

1956 *A History of Education in Antiquity.* New York: Mentor Books. Translation from the French by George Lamb, based on the second edition of this work, Paris: Éditions du Seuil, 1950.

Marx, Karl, and Engles, Friedrich
1947 *Manifesto of the Communist Party.* New York: International Publishers. Authorized translation form the German, edited and annotated by Friedrich Engels of a text originally published in 1848.

1998 *German Ideologie.* Amherst, NY: Prometheus Books. Translation from the German of a work published in 1846.

Matsen, Patricia, Rolinson, Philip, and Souza, Marion
1990 *Readings from Classical Rhetoric.* Carbondale, ILL: Southern Illinois University Press.

Maynard Smith, John
1970 What is the use of sex? *Journal of Theoretical Biology* 30, 319–335.

1972 *On Evolution*. Edinburgh: Edinburgh University Press.

1978 *The Evolution of Sex*. Cambridge: Cambridge University Press.

1982 *Evolution and the Theory of Games*. Cambridge: Cambridge University Press.

1994 *The Theory of Evolution.* Cambridge: Cambridge University Press.

Maynard Smith, John and Harper, David
2003 *Animal Signals.* Oxford: Oxford University Press.

Maynard Smith, John and Price, George R.
1971 The logic of animal conflict. *Nature* 246: 15–18.

Maynard Smith, John and Szathmáry, Eörs
1995 *The Major Transitions in Evolution.* Oxford: W. H. Freeman/Spektrum.

1999 *The Origins of Life: From the Birth of Life to the Origin of Language*. Oxford: Oxford University Press.

Mayr, Ernst
1982 *The Growth of Biological Thought: Diversity, Evolution, and Inheritance.* Cambridge: Harvard University Press.

1990 *One Long Argument: Darwin and the Genesis of Modern Evolutionary Thought*. Cambridge: Harvard University Press.

McCormack, Susan
1998 *The Shadows of Poetry: Virgil in the Mind of Augustine.* Berkeley: The University of California Press.

Meier, Christian
1998 *Athens: A Portrait of the City in the Golden Age*. New York: Henry and Holt. Translation from the German by Robert and Rita Kimber. Original edition, Berlin, Wolf Jobst Siedler, 1993.

Meiggs, Russell
1960 *Roman Ostia.* Oxford: Clarendon Press.

Mesoudi, Alex, Whiten, Andrew, and Laland, Kevin N.
2004 Is human cultural eolution Darwinian? Evidence reviewed from the perspective of *The Origin of Species*. *Evolution* 58 (1), 1–11.

Michod, Robert
1995 *Eros and Evolution: A Natural Philosophy of Sex.* Reading, MA: Addison-Wesley.

Miller, Geoffrey
2000 *The Mating Mind: How Sexual Choice Shaped the Evolution of Human Mind.* New York: Doubleday.

Mirowski, Philip
1991 What were von Neumann and Morgenstern trying to accomplish? In *Toward a History of Game Theory, Annual Supplement volume 24, History of Political Economy,* Riy Weintraug (ed.), 113–147. Durham: Duke University Press.

Nash, Jr., John F.
1997 Reprint. The bargaining problem. In *Classics in Game Theory,* Harold W. Kuhn (ed.), 5–13. Princenton: Princeton University Press. Paper originally Published in *Econometrica* 18 (1950), 155–162.

Nietzsche, Fredrich
1990 *The Twilight of Idols.* Harmondsworth: Penguin Books. Translation from the German by R. J. Hollingdale, with an introduction of Michael Tanner of a work published in 1889.

Nowak, Martin A. and Sigmund, Karl
2004 Evolutionary dynamics of biological games. *Science* 303, 793–799.

Ober, Josiah
1985a *Fortress Attica: Defense of the Athenian Land Frontier: 404–322 BC.* Leiden: Brill.

1985b Thucydides, Pericles, and the strategy of defense. In *The Craft of the Ancient Historian; Essays in Honor of Chester G. Starr*, John W. Eadie and Josiah Ober (eds.), 171–188. Lanhan: University Press of America.University Press.

1998 *Political Dissent in Democratic Athens*. Princeton: Princeton University Press.

O'Donnell, James J.
1992a *Augustine's Confessions I, Introduction and Text*. Oxford: Oxford University Press.

1992b *Augustine's Confessions II, Commentary on Books 1–7.* Oxford: Oxford University Press.

1992c *Augustine's Confessions, Commentary on Books 8–13.* Oxford: Oxford University Press.

Olson, Mancur
1968 *The Logic of Collective Action: Public Good and the Theory of Groups*. New York: Schoken Books.

O'Meara, John J.
1969 *The Young Augusitne; The Growth of Augustine's Mind up to his Conversion.*Staten Island, Y: Alba House.

1991 Augustine's *Confessions*: Elements of fiction. In *Augustine: From Rhetor to Theologian,* John McWilliam, Timothy Barnes, Michael Faye (eds.), 77–96. Waterloo: Wilfrid Laurier.

Owen, Dennis F.
1982 *Camouflage and Mimicry*. Chicago: The Chicago University Press.

Peirce, Charles S.
1931 *The Collected Papers of Charles Pierce,* Charles Hartshorn and Paul Weiss (eds.) Cambridge: Harvard Univeristy Press.

1976 *The New Elements of Mathematics, volume 3: Mathematical Miscellanea,* Carolyn Eisler (ed.). The Hague: Mouton.

1991 *Selected Manuscripts and Letters*. Indianapolis: Peirce Edtion Project. Personal research done during a Fulbright scholarship.

1992 *The Essential Peirce: Selected Philosophical Writings, volume 1 (1867–1893,)* Nathan Houser and Christian Kloesel (eds.). Bloomington: IndianaUniversity Press.

1998 *The Essential Peirce: Selected Philosophical Writings, volume 2 (1893–1913),* Peirce Edition Project (ed.). Bloomington: Indiana University Press.

Perler, Othmar
1971 *Les Voyages de Saint Augustine*. Paris; Études Augustinnienes.

Piddington, Ralph
1957 Malinowski's theory of needs. In *Man and Culture: An Evaluation of the Work of Bronislaw Malinowski*, Raymond Firth (ed.), 33–52. London: Routledge and Kegan Paul.

Plato
1963 *The Collected Dialogues, Including the Letters*, Edith Hamilton and Huntington Cairns (eds.), with introductory and prefatory notes. Princeton: Princeton University Press. Translations from the Greek of works probably written at the beginning of the fourth century B.C.E.

Pleket, Henry Willy

1996 L'agonismo sportive. In *I Greci: Storia, Cultura, Arte Societá, v. 1: oi e I Greci*, Salvatore Sett (ed.), 507–537. Torino: Einaudi.

Plutarch

1967 *The Rise and Fall of Athens: Nine Greek Lives.* London: Folio Society Translation from the Greek,with an introduction and commentary by Ian Scott- Kilvert, of works written at the end of the first and beginning of the second century CE.

Polanyi, Karl

1968 *Primitive, Archaic, and Modern Economies.* Garden City, NY: Doubleday.

1975 *The Great Transformation.* New York: Norton.

Poole, J F. C., and Holladay A. J.

1979 Thucydides and the plague of Athens. *Classical Quarterly* 29, 282–310.

Popper, Karl R.

1963 Reprint. *The Open Society and its Enemies, volume 2: The High Tide* of Propecy: Hegel, Marx, and the Aftermath. New York: Harper Torch. Original edition, London, Routledge and Kegan Paul, 1945.

Poundstone, William

1992 *The Prisoner's Dilemma.* New York: Anchor Books.

Powell, E. C. and Jones, C. E.

1983 Floral mutualism in *Lupinus Benthamii* (*Fabacea*) and *Delphinium Parryl (Ranunculaceae).* In *Handbook of Experimental Pollination Biology,* C. E. Jones and R. J. Little (eds.), 310–329. New York: Scientific and Academic Editions.

Pradeau, Jean-François
2004 L'ébriété democratique: La critique platonicienne de la démocratie dans *Les Lois*. *The Journal of Hellenic Studies* 124, 108–124.

Quine, W. V.
1969 Foreword. In *Convention: A Philosophical Study,* David Lewis, x-xii. Cambridge: Harvard University Press.

Radcliffe-Brown A. R.
1958 *Methods in Social Antrhopology*. Chicago: The University of Chicago Press.

Ramachandran, V. S., and Blakeslee, Sandra
1998 *Phantoms in the Brain: Probing the Mysteries of the Human Mind.* New York: Morrow.

Ridley, Matt
1994 *The Red Queen: Sex and the Evolution of Human Nature.* Harmondsworth: Penguin.

2003 *Nature via Nurture; Genes, Experience, and What Makes us Humans* New . York: Harper/Collins.

Riechert, Susan E. and Hammerstein, Peter
1983 Game theory in the ecological context. *Annual Review of Ecology and Systematics* 14, 377–409.

Robinson, Richard
1995 The theory of names in Plato's *Cratylus*. *Revue Internationale de Philosophie* 9, 221–236.

Romilly, Jacqueline de
1963 *Thucydides and the Athenian Imperialism.* Oxford: Basil Blackwell. Translation from the French by Philp Thody. Original edition, Paris, Societé d'Édition "Les Belles Lettres, " 1947.

1967 *Histoire et Raison chez Thucydides*. Paris: Societé d'Études "Les BellesLettres."

1975 *Magic and Rhetoric in Ancient Greece*. Cambridge: Harvard University Press.

1991 *The Great Sophists in the Periclean Age*. Oxford: Clarendon Press. Translation from the French by Janet Lloyd. Original edition, Paris: Éditions de Fallois, 1988.

Rood, Tim
2002 Thucydides in the Polis. *The Classical Review* 51, 241–242.

Rousseau, Jean-Jacques
1994 *Social Contract*. Hanover, NH: University Press of New England. Translation from the French by Judith R. Bush, Roger D. Masters and Christopher Kelly (eds.) of a work originally published in 1762.

Russell, Bertrand
1968 Reprint. *Common Sense and the Nuclear Warfare.* New York: AMS. Original edition, New York: Simon and Schuster, 1959.

Sahlins, Marshall
1976 *Culture and Practical Reason.* Chicago: The University of Chicago Press.

1979 *The Use and Abuse of Biology: An Anthropological Criticism of Sociobiology*: Ann Arbour: The University of Michigan Press.

2004 *Apologies to Thucydides: Understanding History as Culture and Vice Versa.*Chicago: The University of Chicago Press.

Saussure, Ferdinand de
1916 *Cours de Linguistique Génerale*, Charles Bally, Alfred Sechehaye (eds.) with the collaboration of A. Reindlinger. Paris: Payot.

Schaller, George B.
1972 *The Serengeti Lion: A Study of Predator-Prey Relations.* Chicago: The University of Chicago Press.

Schelling, Thomas C.
1960 *The Strategy of Conflict*. Cambridge: Harvard University Press.

Segerstråle, Ullica
2000 *The Defenders of the Truth; the Battle for Science in the Sociobiological Debate and Beyond.* New York: Oxford University Press.

Selten, Reinhard, and Kirschiker, Wilhem
1982 Comparison of two theories of characteristic function experiments. In *Aspiration Levels in Bargaining and Economic Decision Making, Proceedings of the Third Conference on Experimental Economics, Wizenhold, Germany, August 29–Setember 3, 1982,* Rienhard Tietz (ed.), 258–264. Berlin: Springer Verlag.

Sen, Amartya
1969 A game-theoretic analysis of theories of collectivism in allocation. In *Growth and Choice,* Tapas Majumdar (ed.), 1–17. Oxford: Oxford University Press.

1977 Rational fools: A critique of the behavioral foundations of economic theory. *Philosophy and Public Afffairs* 6, 317-344.

1997 *Choice, Welfare, and Measurement.* Cambridge: Cambridge University Press.

Shorey, Paul
1968 *The Unity of Plato's Thought.* Hamden, CT: Archon.

Sigmund, Karl
1995 Reprint. *Games of Life: Explorations in Ecology, Evolution, and Behaviour*. Harmondsworth: Penguin. Original edition, Oxford: Oxford University Press, 1993.

Singer, Peter
1999 *A Darwinian Left: Politics, Evolution, and Cooperation.* New Haven: Yale University Press.

Spence, I. G.
1990 Perikles and the defense of Attica during the Peloponnesian War. *Journal of Hellenic Studies* 60, 9–19.

Spivey, Nigel
2004 *The Ancient Olympics: A History.* Oxford: Oxford University Press.

Sprague, Rosamund Kent (ed.)
1971 *The Older Sophists: A Complete Translation of Several Authors of the Fragments on Die Fragmente der Vorksocratiker, edited by Dieter Kranz.* Columbia, SC: University of Souht Carolina Press.

Solomon, John
1989 Thucydides and the recognition of contagion. *Maya* 37, 122–123.

Starnes, Colin
1992 Augustine's conversion and the ninth book of the *Confessions.* In *Augustine: From Rhetor to Theologian,* John McWilliam, Timothy Barnes, Michael Fahey, Peter Slater (eds.), 51–66. Waterloo: Wilfried Laurier Press.

Stock, Brian
1996 *Augustine, the Reader: Meditation, Self-Knowledge, and the Ethics of Interpretation.* Cambridge: Harvard University Press.

Stocking, George W. Jr.
1983 An ethnographer's magic: Fieldwork in British anthropology from Tylor to Malinowski. In *History of Anthropology I,* 70–120. Madison, WI: University of Wisconsin Press.

Testard, Maurice
1958 *Augustine et Cicéron, volume 1: Cicéron dans la formation et l'oeuvre de Saint Augustine.* Paris: Études Augustinniebes.

Thucydides
1972 *The History of the Peloponnesian War.* Harmondsworth: Penguin Books. Translation from the Greek by Rex Warner,

with an introduction Of M. I. Finley of a work probably left incomplete in 404 B.C.E.

Thom, René
1975 *Structural Stability and Morphogenesis; An Outline of a General Theory of Models.* Reading, MA: W. A. Benjamin. Translated from the French by D. H. Fowler, with a foreword by C. H. Waddington. Original edition, Paris, 1972.

1983 *Mathematical Models of Morphogenesis.* Chischester: Ellis Horwood. Translation from the French by W. M. Brookes and D. Rand. Original edition, Paris: Christian Bourgois, 1980.

Tinbergen, Nikko
1972-3 *The Animal and its World: Explorations of an Ethologist, 1932–1972.* Cambridge: Harvard University Press.

Tocqueville, Alexis
1994 *Democracy in America.* New York: Alfred Knopf. Translated from the French by Francis Bowen, revised and updated, with an introduction by Alan Ryan of a work published in two parts in 1835 and 1840.

Tregenza, Tom
2003 The battle of sexes. *Nature* 423, 929–930.

Trivers, Robert
1971 The evolution of reciprocal altruism. *The Quarterly Review of Biology* 46: 35–57.

1985 *Social Evolution.* Menlo Park, CA: Benjamins/Cummings.

Troubetzkoy, Nikolai
2001 *Studies in General Linguistics and Language Structure* Durham, NC: Duke University Press. Translation from the Russian by Marvin Taylor and Anatoly Liberman of papers edited by Anatoly Liberman.

Tucker, A. W.
1980 On jargon: The prisoner's dilemma. *UMAP Journal* 1, 101.

Tylor, E. B.
1871 *Primitive Culture: Researches into Developments of Mythology, Philosophy, Religion, Art, and Custom.* London: J. Murray.

van Valen, Leigh
1973 A new evolutionary law. *Evolution Theory* 1, 1–30.

von Neumann, John, and Morgenstern, Oskar
1944 *Theory of Games and Economic Behavior.* Princeton: Princeton University Press.

Weingartener, Rudolph H.
1972 Making sense of the *Cratylus. Phronesis* 15, 5–25.

Westlake, H. D.
1969 *Individuals in Thucydides.* Princeton : Princeton University Press.

Whiten, Andrew, Horner, Victoria, and Marshall-Pescini, Sarah
2003 Cultural pantanthropology. *Evolutionary Anthropology* 12, 92–105.

Whitney, W. D.
1875 *The Life and Growth of Languages: An Outline of Linguistic Science.* New YorK: D. Appleton.

Wickler, Wolfgang
1968 *Mimicry in Plants and Animals.* New York: McGraw-Hill. Translation from the German by R. D. Martin. Original edition, 1968

Willey, R. Haven
1983 The evolution of communication: information and manipulation. In *Communication (Animal Behavior), volume 2,* T. K. Halliday and P. J. B. Slater, 156–189. Oxford: Basil Blackwell.

1996 Errors, exaggeration, and deception in animal communication. In *Behavioral Mechanisms in Evolutionary Ecology*, Leslie A.

Real (ed.), 157–189. Chicago: The University of Chicago Press.

Williams, Bernard
2001 *Truth and Truthfulness.* Princeton: Princeton University Press.

Williams, George C.
1964 Measurements of consotiation among fish and comments on the evolution of schooling. *Publications of the Museum, Michigan State University, Biological Series* 2/7: 149–184.

1974 *Sex and Evolution.* Princeton: Princeton University Press.

1992 *Natural Selection: Domains, Levels, and Challenges.* New York: Oxford University Press.

1996 *Adaptation and Natural Selection: A Critique of Some Current Evolutionary Thought.* Princeton: Princeton University Press.

1997 *The Pony's Fish Glow and Other Clues to Plan and Purpose in Nature.* New York: Basic Books.

Wilson, E. O.
1968 Chemical systems. In *Animal Communication: Techniques of Study and Results of Research,* T. A. Sebeok (ed.) 75–102. Bloomington: Indiana University Press.

1975 *Sociobiology: The New Synthesis.* Cambridge: The Belknap Press of Harvard University Press.

Wright, Robert
2000 *Nonzero: The Logic of Human Destiny.* New York: Pantheon.

Wrong, Dennis H.
1961 The oversocialized concenption of Man in modern sociology. *American Sociological Review* 26(2), 183–193.

Wu, Jianzhong, and Axelrod, Robert
1995 How to cope with noise in the iterated Prisoner's Dilemma. *Journal of Conflict Resoution* 19 (1), 183–189. Reprinted in Axelrod (1997.)

Xenophon
1910 Memorabilia of Socrates. In *Socratic Discourses by Plato and Xenophon*. London: J. M. Dent. Translation from the Greek by Carnes Lloyd of a work written around 400 B.C.E.

Yalman, N.
1969 De Tocqueville in India: An essay on the caste system. *Man* 4: 121–131.

Yong-Hoi, Choi
2002 *Le Problème du Temps chez Ferdinand de Saussure*. Paris: L'Harmattan.

Zahavi, Amotz
1980 Ritualization and the evolution of movement signals. *Behavior* 72, 77-81

1993 The fallacy of conventional signaling. *Philosophical Transactions of the Royal Society of London* 340, 227–230.

Zahavi, Amotz and Zahavi, Avishiga
1997 *The Handicap Principle: A Missing Piece of Darwin's Puzzle*. New York: Oxford University press.

Zokolwski, John E.
1985 *Thucydides and the Tradition of Funeral Speeches in Athens*. Salem, NH: Ayer.

Zuk, Marlene
2000 *Sexual Selection: What we can and can't learn about Sex from Animals*. Berkeley: The University of California Press.

Index

Bei Fragen zur Produktsicherheit wenden Sie sich bitte an:
If you have any questions regarding product safety,
please contact:

Walter de Gruyter GmbH
Genthiner Straße 13
10785 Berlin
productsafety@degruyterbrill.com